*To Ivana, Ondřej, and Frances*
*without whose support and encouragement*
*this would have forever remained a work in progress.*

Miroslav Bartušek
Zuzana Došlá
John R. Graef

# The Nonlinear
# Limit–Point/Limit–Circle
# Problem

Birkhäuser
Boston • Basel • Berlin

Miroslav Bartušek
Department of Mathematics
Masaryk University
601 77 Brno
Czech Republic

Zuzana Došlá
Department of Mathematics
Masaryk University
601 77 Brno
Czech Republic

John R. Graef
Department of Mathematics
University of Tennessee
Chattanooga, TN 37403
U.S.A.

**Library of Congress Cataloging-in-Publication Data**

A CIP catalogue record for this book is available from the Library of Congress,
Washington D.C., USA.

AMS Subject Classifications: 34B, 34C, 34D, 34E, 34L

ISBN 0-8176-3562-9          Printed on acid-free paper.

©2004 Birkhäuser Boston          *Birkhäuser*

Printed in the United States of America.     (MV)

9 8 7 6 5 4 3 2 1          SPIN 10959886

Birkhäuser is part of *Springer Science+Business Media*

*www.birkhauser.com*

# Preface

The purpose of this book is to present some new developments in the asymptotic analysis of nonlinear differential equations with particular attention paid to the limit–point/limit–circle problem. Nearly one hundred years have passed since Hermann Weyl first investigated this problem for second order linear differential equations. Since then this problem has been extended in various directions including:

- spectral analysis of differential operators of order $2n$;

- LP/LC problem for second order nonlinear differential equations;

- LP/LC problem for higher order nonlinear differential equations;

- the problem of existence of solutions not in $L^2$ for second order equations;

- the problem of existence of at least one solution in $L^2$ for second and higher order equations;

- the problem of existence of at least one solution in $L^p$ for second and higher order equations.

Our attention here is focused on the extension of the classical Weyl problem to nonlinear equations in the sense that either all solutions are of the "nonlinear limit–circle type" or there is at least one solution that does not have this property. Some related problems, such as the existence of an $L^2$ solution, are not treated here. We should emphasize that for nonlinear problems, the existence of continuable solutions and singular solutions plays an important role. (This is discussed in detail in Chapter 2.)

The book consists of nine chapters. Chapter 1 discusses the origin of the limit–point/limit–circle problem including the motivation for the choice of this terminology. Chapter 2 gives the basic definitions and extension for nonlinear differential equations and examines the question of the existence of both continuable and singular solutions. Chapter 3 presents our results for second order nonlinear equations, including some necessary and sufficient conditions for a second order nonlinear equation to be of the "nonlinear limit–circle type." Chapter 4 describes some early attempts at obtaining limit–point type results for second and higher order nonlinear equations. In the last section in this chapter, we also describe some recently obtained results that are related to these earlier ones. In Chapter 5, we examine the connection between the limit–circle property and other properties of solutions of linear and nonlinear equations such as boundedness, oscillation, and convergence to zero. Chapters 6 and 7 examine the limit–point/limit–circle problem for third and fourth order equations, respectively. Chapter 8 is devoted to equations of arbitrary order, and Chapter 9 discusses the relationship between the limit–point/limit–circle problem and the spectral theory of differential operators. There are more than 120 references, and a number of open problems for future research are included.

Our joint interest in this problem began in the fall of 1993 when J. R. Graef visited Brno and gave a survey lecture on the status of the nonlinear limit–point/limit–circle problem. His own interest in the problem began in the late 1970s and included some collaboration with P. W. Spikes on second order nonlinear equations while they were both on the faculty at Mississippi State University. With that initial visit to Brno, the present authors began collaborating on the nonlinear limit–point/limit–circle problem and this led quite naturally to the present monograph.

We wish to express our thanks to Doc. RNDr. Jaromir Kuben, CSc., for his assistance in preparing the electronic files for this manuscript. We also wish to thank Ann Kostant and the staff at Birkhäuser Boston; they are an excellent team to work with.

M. Bartušek

Z. Došlá

J. R. Graef

Brno, Czech Republic and Chattanooga, Tennessee

October 2003

# Contents

# Basic Notation

| | |
|---|---|
| $\mathbb{R}$ | the set of real numbers; |
| $\mathbb{C}$ | the set of complex numbers; |
| $\mathbb{R}_+$ | the set of nonnegative real numbers, i.e., the interval $[0, \infty)$; |
| $\mathbb{R}^n$ | the set $\mathbb{R} \times \cdots \times \mathbb{R}$ ($n$ times); |
| $L^2$ | the set of Lebesgue square integrable functions $u: \mathbb{R}_+ \to \mathbb{R}$; |
| $L([a, b])$ | the set of Lebesgue integrable functions $u: [a, b] \to \mathbb{R}$; |
| $L_{loc}(\mathbb{R}_+)$ | the set of integrable functions $u: \mathbb{R}_+ \to \mathbb{R}$ whose restriction to any interval $[a, b]$ belongs to $L([a, b])$; |
| $L^2_{loc}(\mathbb{R}_+)$ | the set of locally square integrable functions on $[0, \infty)$; |
| $AC_{loc}(\mathbb{R}_+)$ | set of all locally absolutely continuous functions on $[0, \infty)$; |
| $C^k([a, b])$ | the set of $k$ times continuously differentiable functions $u: [a, b] \to \mathbb{R}$; |
| $C^0(\mathbb{R}_+)$ | the set of continuous functions $u: \mathbb{R}_+ \to \mathbb{R}$; |
| $g(t) = \mathcal{O}(h(t))$ | $g(t)/h(t)$ is bounded as $t \to \infty$; |
| $[\![n]\!]$ | the greatest integer function of $n$. |

# Chapter 1

# Origins of the Limit–Point/ Limit–Circle Problem

In this chapter, we begin with a discussion of the origins of the limit–point/limit–circle problem including a motivation for the choice of this terminology. We then discuss its relationship to the notion of the deficiency index and describe the classical results for second order linear equations.

## 1.1. The Weyl Alternative

In 1910, Hermann Weyl [114] published his now classic paper on eigenvalue problems for second order linear differential equations of the form

$$(a(t)y')' + r(t)y = \lambda y, \quad t \in [0, \infty), \ \lambda \in \mathbb{C}, \tag{1.1}$$

and he classified this equation to be of the *limit–circle* type if every solution is square integrable, i.e., belongs to $L^2$, and to be of the *limit–point* type if at least one solution does not belong to $L^2$. In the ensuing years there has been a great deal of interest in the limit–point/limit–circle problem due to its importance in relation to the solution of certain boundary value problems (see Titchmarsh [109, 110]). As we will see later, the study of the analogous problem for nonlinear equations is relatively new and not nearly as extensive as for the linear case.

To understand the basis for Weyl's terminology, we begin with one of his fundamental results. The terminology limit–point/limit–circle arises in a somewhat natural way from the proof of this result, a sketch of which will be given.

**Theorem 1.1.** *If Im $\lambda \neq 0$, then (1.1) always has a solution $y \in L^2(\mathbb{R}_+)$, i.e.,*

$$\int_0^\infty |y(t)|^2 dt < \infty.$$

*Sketch of the Proof.* For $\lambda$ with $\text{Im }\lambda \neq 0$, let $\varphi$ and $\psi$ be two linearly independent solutions of (1.1) satisfying the initial conditions

$$\varphi(0, \lambda) = 1, \quad \psi(0, \lambda) = 0,$$
$$\varphi'(0, \lambda) = 0, \quad \psi'(0, \lambda) = 1.$$

The functions $\varphi(t, \lambda)$ and $\psi(t, \lambda)$ are analytic in $\lambda$ on $\mathbb{C}$. Then, any other solution $y$ is a linear combination of these solutions, say,

$$y(t, \lambda) = \varphi(t, \lambda) + m(\lambda)\psi(t, \lambda).$$

Choose $b > 0$ and let $c_1$ and $c_2$ be arbitrary but fixed constants; we want to determine $m(\lambda)$ so that the solution $y$ satisfies

$$c_1 y(b, \lambda) + c_2 y'(b, \lambda) = 0. \tag{1.2}$$

This desired value of $m$ depends on $\lambda$, $b$, $c_1$, and $c_2$, and in fact has the form of the linear fractional transformation

$$m = \frac{Az + B}{Cz + D}.$$

The image of the real axis in the $z$-plane is a circle $\mathcal{C}_b$ in the $m$-plane. The solution $y$ will satisfy (1.2) if and only if $m$ is on $\mathcal{C}_b$. An argument using Green's identity shows that this is true if and only if

$$\int_0^b |y(s)|^2 ds = \frac{\text{Im }m}{\text{Im }\lambda},$$

and the radius of the circle $\mathcal{C}_b$ is

$$r_b = \left( 2 \text{Im }\lambda \int_0^b |y(s)|^2 ds \right)^{-1} \tag{1.3}$$

Observe that if $b_1 < b$, then

$$\int_0^{b_1} |y(s)|^2 ds < \int_0^b |y(s)|^2 ds,$$

so $r_b < r_{b_1}$, i.e., the circle $\mathcal{C}_{b_1}$ contains the circle $\mathcal{C}_b$ in its interior. Thus, as $b \to \infty$, the circles $\mathcal{C}_b$ converge either to a circle $\mathcal{C}_\infty$ or to a point $m_\infty$. If the limiting form is a circle, then $r_\infty > 0$, and so (1.3) implies

$$\int_0^\infty |y(s)|^2 ds < \infty,$$

i.e., $y \in L^2$ for any $m$ on $\mathcal{C}_\infty$. If the limit is the point $m_\infty$, then $r_\infty = 0$ and there is only one solution in $L^2$. $\qquad\square$

Titchmarsh [109, 110] discusses the connection between the limit–point property and the existence of a unique Green's function for second order linear differential equations. In the limit–circle case, the Green's function depends on a parameter.

Essential to the study of the limit–point/limit–circle problem is the following result of Weyl.

**Theorem 1.2.** *If* (1.1) *is limit-circle for some* $\lambda_0 \in \mathbb{C}$, *then* (1.1) *is limit-circle for all* $\lambda \in \mathbb{C}$.

In particular, Theorem 1.2 holds for $\lambda = 0$, that is, if we can show that equation (1.1) is limit–circle for $\lambda = 0$, then it is limit–circle for all values of $\lambda$. Moreover, if (1.1) is not limit–circle for $\lambda = 0$, then it is not limit–circle for any value of $\lambda$. In view of Theorem 1.1, for second order equations the problem reduces to whether equation (1.1) with $\operatorname{Im}\lambda \neq 0$ has one (limit–point case) or two (limit–circle case) solutions in $L^2$ (this is known as the *Weyl Alternative*). As we will see later, the situation for higher order equations is somewhat different in that the limit–point and limit–circle cases do not form a dichotomy.

The limit–point/limit–circle problem then becomes that of determining necessary and/or sufficient conditions on the coefficient functions to be able to distinguish between these two cases. Weyl's results have spawned research in a variety of directions including the study of what is called the *deficiency index problem*, which we describe in the next section.

## 1.2. The Deficiency Index Problem

The extension of the limit–point/limit–circle problem for second order equations to equations of higher order leads to the study of the deficiency index for self-adjoint differential operators. Consider the differential expression

$$\ell(y) \equiv \sum_{i=0}^{n} (-1)^i \left( p_i(t) y^{(i)} \right)^{(i)} = (-1)^n (p_n(t) y^{(n)})^{(n)}$$
$$+ (-1)^{n-1}(p_{n-1}(t) y^{(n-1)})^{(n-1)} + \cdots + (-1)(p_1(t) y^{(1)})^{(1)} + p_0(t) y \quad (1.4)$$

where $p_i$, $i = 0, \ldots, n$ are real-valued functions for $t \in \mathbb{R}_+$, $p_n(t) > 0$, and $p_n^{-1}, p_{n-1}, \ldots, p_0 \in L_{\text{loc}}(\mathbb{R}_+)$. Then, the minimal operator $L_0$ associated with this differential expression has self-adjoint extensions; see, for example, Naimark

[94, §17]. The *deficiency index*, denoted by $m$, of $L_0$ is the number of linearly independent solutions of

$$\ell(y) = \lambda y, \quad \text{Im} \lambda \neq 0, \tag{1.5}$$

that are in $L^2$ (see Devinatz [34]). The number $m$ is independent of $\lambda$ (as long as $\text{Im} \lambda \neq 0$) and the possible values for $m$ are

$$n, n + 1, \ldots, 2n,$$

with the exact value taken depending on the coefficients $p_i$.

There is a higher order counterpart of Theorem 1.2 above (see Naimark [94, Theorem 4, p. 93]). That is, (1.5) has all its solutions in $L^2$, i.e., is (higher order) limit-circle if and only if the equation

$$\ell(y) = 0 \tag{1.6}$$

has all its solutions in $L^2$. Thus, in what follows, we only study the limit–circle problem for equation (1.6).

For some time it was believed that the deficiency index was always either $n$ or $2n$, and as we saw in Section 1.1, this is the case for second order equations. However, for higher order equations, any value between $n$ and $2n$ is possible (see Glazman [57]). Some authors refer to the case $m = n$ as the limit–point case for higher order linear equations.

Several conjectures on the value of the deficiency index $m$ have been posed over the years; here we will briefly discuss a couple of them.

• Everitt's conjecture (1961): *If $p_j \geq 0$ for all $j = 0, \ldots, n$, then $m = n$.*

• Kauffman (1976) disproved this conjecture giving the example that the operator $-(x^a y''')''' + K x^{6-a} y$ has the deficiency index $m > 3$.

• McLeod's conjecture (1962): *If $p_j \geq 0$ for all $j = 0, \ldots, n$, then $n \leq m \leq 2n - 1$ and all $m$ occur.*

As far as we know this conjecture is still open.

• Paris and Wood (1981) proved: *For every $j$ with $0 \leq j \leq [\![ \frac{n+1}{4} ]\!]$ there exists a real formally self-adjoint expression of order $2n$ with nonnegative coefficients having deficiency index $m = n + 2j$.*

• Schultze (1992) improved the range of values covered by showing: *For every $j$, $0 \leq j < n/2$ there exists such an expression having deficiency index $m = n + 2j$.*

This still leaves half of the values between $n$ and $2n$ unaccounted for.

An excellent historical account of the development of the deficiency index problem can be found in the survey article by Everitt [49] which contains more than sixty references to work prior to 1976; also see [50, 51].

## 1.3. Second Order Linear Equations

In the study of the linear equation

$$(a(t)y')' + r(t)y = 0, \tag{1.7}$$

where $a, r : \mathbb{R}_+ \to \mathbb{R}$ is continuous, $a', r' \in AC_{\text{loc}}(\mathbb{R}_+)$, $a'', r'' \in L^2_{\text{loc}}(\mathbb{R}_+)$, $a(t) > 0$, and $r(t) > 0$, it has proven useful (see Dunford and Schwartz [37] or Burton and Patula [24]) to make the transformation

$$s = \int_0^t \left[ \frac{r(u)}{a(u)} \right]^{\frac{1}{2}} du, \quad x(s) = y(t), \tag{1.8}$$

and let "$\cdot$" denote $\frac{d}{ds}$. Then, we have

$$y'(t) = \dot{x}(s)\frac{ds}{dt} = [r(t)/a(t)]^{\frac{1}{2}}\dot{x}(s),$$

$$a(t)y'(t) = [a(t)r(t)]^{\frac{1}{2}}\dot{x}(s),$$

and

$$(a(t)y'(t))' = [a(t)r(t)]^{\frac{1}{2}}\ddot{x}(s)[r(t)/a(t)]^{\frac{1}{2}} + \frac{1}{2}[a(t)r(t)]^{-\frac{1}{2}}[a(t)r(t)]'\dot{x}(s)$$

$$= r(t)\ddot{x}(s) + [a(t)r(t)]'\dot{x}(s)/2[a(t)r(t)]^{\frac{1}{2}}.$$

Equation (1.7) then becomes

$$\ddot{x}(s) + 2p(t)\dot{x}(s) + x(s) = 0$$

where

$$p(t) = [a(t)r(t)]'/4a^{\frac{1}{2}}(t)r^{\frac{3}{2}}(t).$$

The following theorem due to Dunford and Schwartz [37, p. 1410] is probably the best known limit–circle result for equation (1.7).

**Theorem 1.3.** *Assume that*

$$\int_0^\infty \left| \left[ \frac{(a(u)r(u))'}{a^{\frac{1}{2}}(u)r^{\frac{3}{2}}(u)} \right]' + \frac{\{[a(u)r(u)]'\}^2}{4a^{\frac{3}{2}}(u)r^{\frac{5}{2}}(u)} \right| du < \infty. \tag{1.9}$$

*If*

$$\int_0^\infty [1/(a(u)r(u))^{\frac{1}{2}}]du < \infty, \tag{1.10}$$

*then equation (1.7) is limit–circle, i.e., every solution $y(t)$ of (1.7) satisfies*

$$\int_0^\infty y^2(u)du < \infty.$$

Their corresponding limit–point result is the following.

**Theorem 1.4.** *Assume that (1.9) holds. If*

$$\int_0^\infty [1/(a(u)r(u))^{\frac{1}{2}}]du = \infty, \tag{1.11}$$

*then equation (1.7) is limit–point, i.e., there is a solution $y(t)$ of (1.7) such that*

$$\int_0^\infty y^2(u)du = \infty.$$

**Remark 1.1.** Everitt [47] proved that the linear equation (1.7) is of the limit–circle type if (1.10) holds and the condition (1.9) of Dunford and Schwartz is replaced by

$$\int_0^\infty \{[a(u)(a(u)r(u))^{-\frac{5}{4}}(a(u)r(u))']'\}^2 du < \infty. \tag{1.12}$$

Wong [118, Proposition, p. 424] showed that equation (1.7) is of the limit–circle type if (1.10) holds and

$$\int_0^\infty \left| [a(u)r(u)]^{-\frac{1}{4}}\{a(u)[(a(u)r(u))^{-\frac{1}{4}}]'\}' \right| du < \infty. \tag{1.13}$$

By showing that conditions (1.10) and (1.12) imply (1.13), he thus has an extension of Everitt's result.

**Remark 1.2.** When $a(t) \equiv 1$ so that equation (1.7) reduces to

$$y'' + r(t)y = 0, \tag{1.14}$$

then condition (1.9) of Dunford and Schwartz becomes

$$\int_0^\infty \left| \frac{r''(u)}{r^{3/2}(u)} - \frac{5}{4}\frac{[r'(u)]^2}{r^{5/2}(u)} \right| du < \infty. \tag{1.15}$$

Burton and Patula [24, Theorem 1] (also see Knowles [84, Theorem 5]) proved a variation of Theorem 1.3 for equation (1.14) by replacing conditions (1.9) and (1.10) (with $a(t) \equiv 1$) of Dunford and Schwartz with the single condition

$$\int_0^\infty \left[ \frac{r(0)}{r(s)} \right]^{1/2} \exp\left\{ (1/4) \int_0^s \left| \frac{(r'(u))^2}{4r^{\frac{5}{2}}(u)} + \left[ \frac{r'(u)}{r^{\frac{3}{2}}(u)} \right]' \right| du \right\} ds < \infty. \quad (1.16)$$

Knowles [82] showed that conditions like (1.9) with (1.10), or (1.12) with (1.10), are special cases of a broader class of conditions for linear equations to be of the limit–circle type. For example, Knowles shows that a result of Pavljuk [100], namely, equation (1.14) is of the limit–circle type provided (1.10) holds and

$$\left| \frac{r''(t)}{r(t)} - \frac{5}{4} \left( \frac{r'(t)}{r(t)} \right)^2 \right| \quad \text{is bounded,} \quad (1.17)$$

is also in this family of conditions.

Ráb [101, Section 2.3] obtained some asymptotic formulas for solutions of equation (1.14) under the assumption that

$$\int_0^\infty \left| \frac{r''(u)}{r^{3/2}(u)} - \eta \frac{[r'(u)]^2}{r^{5/2}(u)} \right| du < \infty \quad (1.18)$$

with $\eta \in \mathbb{R} - \{\frac{3}{2}\}$.

Harris [71] also studied the limit–circle problem for equation (1.7) under conditions similar to those described above.

Titchmarsh [109, Theorem 5.11] or [110, Section 3] proved the following limit–circle result.

**Theorem 1.5.** *If $r' > 0$, $r(t) \to +\infty$ as $t \to \infty$, $r''$ is eventually of one sign, $r'(t) = \mathcal{O}(|r(t)|^c)$ as $t \to \infty$ for $0 < c < 3/2$, and*

$$\int_0^\infty \frac{1}{r^{\frac{1}{2}}(u)} du < \infty,$$

*then equation (1.14) is of the limit–circle case.*

The following lemma due to Coppel [31] provides some insight into the relationship between conditions such as (1.9), (1.12), (1.13), (1.15), (1.16), and (1.18).

**Lemma 1.1.** ([31, p. 121, Lemma 6]) *If $h \in L^2_{loc}[0, \infty)$ is a positive function such that*

$$\int_0^\infty \left| \frac{h''(u)}{h^{3/2}(u)} - \varepsilon \frac{[h'(u)]^2}{h^{5/2}(u)} \right| du < \infty$$

*for some $\varepsilon \neq \frac{3}{2}$, then the following three conditions are equivalent:*

*(i)* $\displaystyle\int_0^\infty \frac{[h'(u)]^2}{h^{5/2}(u)} du < \infty;$

*(ii)* $\dfrac{h'(t)}{h^{3/2}(t)} \to 0$ *as $t \to \infty;$*

*(iii)* $\displaystyle\int_0^\infty h^{1/2}(u) du = \infty.$

*Moreover, all three of these conditions hold if $\varepsilon < 1$ or $\varepsilon > \frac{3}{2}$.*

The conditions

$$\int_0^\infty h^{1/2}(u) du = \infty \quad \text{and} \quad \int_0^\infty |h(u)|^{-\frac{1}{4}} [(h(u))^{-\frac{1}{4}}]'' | du < \infty$$

taken together are equivalent to

$$\int_0^\infty |h^{-\frac{3}{2}}(u) h''(u)| du < \infty.$$

This can be seen from the above lemma with $\varepsilon = 0$ and $\varepsilon = \frac{5}{4}$.

The connection between the limit–circle property and other asymptotic properties of solutions such as the boundedness, oscillation, and convergence to zero will be discussed in Chapter 5.

The first known limit–point result is actually due to Weyl himself.

**Theorem 1.6.** *If $r(t) \leq r_0$, then (1.14) is of the limit–point type.*

Hartman and Wintner improved Weyl's result as follows.

**Theorem 1.7.** ([74, p. 206]) *If*

$$r(t) \leq ct^2$$

*for all large $t$ and some $c \geq 0$, or more generally, if*

$$\int_0^t r(u) du = \mathcal{O}(t^3) \text{ as } t \to \infty,$$

*then (1.14) is of the limit–point type.*

Hartman and Wintner [74, p. 207] go on to show that the integral condition in Theorem 1.7 is the best possible in the sense that if $r(t) = t^{2+\varepsilon}$ in equation (1.14), then all solutions belong to $L^2$. Hartman and Wintner also proved the following two rather interesting limit–point results.

**Theorem 1.8.** ([73, Section 3]) *If $r$ is positive, nondecreasing, and*

$$\int_0^\infty \frac{1}{r^{\frac{1}{2}}(u)} du = \infty,$$

*then equation (1.14) is of the limit–point type.*

**Theorem 1.9.** ([73, Section 8]) *If $r$ is positive, monotonic, and either*

$$\int_0^\infty \frac{1}{r(u)} du = \infty$$

*or there is a sequence $\{t_n\} \to \infty$ as $t \to \infty$ such that*

$$\limsup_{t\to\infty} \frac{r(t_n)}{t_n} < \infty,$$

*then equation (1.14) has no nontrivial solutions belonging to $L^2$.*

In that same paper, Hartman and Wintner also give a variation on Theorem 1.8 that involves the condition

$$\limsup_{t\to\infty} \frac{r'(t)}{r^{\frac{3}{2}}(t)} < \infty.$$

The following theorem is a well-known limit–point result due to Levinson [92, Theorem IV].

**Theorem 1.10.** *If there is a positive nondecreasing function $R$ such that*

$$r(t) \leq R(t),$$

$$\int_0^\infty \frac{1}{[a(u)R(u)]^{\frac{1}{2}}} du = \infty,$$

*and*

$$\limsup_{t\to\infty} \frac{a^{\frac{1}{2}}(t)R'(t)}{R^{\frac{3}{2}}(t)} < \infty, \tag{1.19}$$

*then equation (1.7) is of the limit–point type.*

Note the similarity between some of the hypotheses in Theorem 1.10 and those in Theorem 1.5. Also see Hartman and Wintner [73, Section 7].

Coddington and Levinson [30, p. 231, Corollary 1] extended Theorem 1.6 to equation (1.7) under the assumption

$$\int_0^\infty \frac{1}{a^{\frac{1}{2}}(u)} du = \infty. \tag{1.20}$$

Everitt [45] was able to drop condition (1.20). In particular, he proved that equation (1.7) is of the limit–point type if $r$ is bounded above and

$$\int_0^\infty \frac{1}{a^{\frac{1}{2}}(u)} du < \infty.$$

A short proof of Everitt's result is given by Wong [117]

Wong and Zettl proved the following interesting limit–point result.

**Theorem 1.11.** ([121, Corollary 4]) *If*

$$r(t) \leq t^{\frac{1}{2}}(a(t)t^{-\frac{3}{2}})'/2, \tag{1.21}$$

*then equation* (1.7) *is of the limit–point type.*

Other limit–point criteria for linear equations include a result of Friedrichs [55] who proved that equation (1.7) is of the limit–point type if $r(t)$ is bounded above and

$$\int_0^\infty \frac{1}{a^2(u)} du = \infty.$$

Hinton [75, p. 175] gives the following example to aid in understanding the relationship between the functions $a$ and $r$.

**Example 1.1.** Consider the equation

$$(t^\alpha y')' + r(t)y = 0, \quad t \geq 1, \tag{1.22}$$

where $r : \mathbb{R}_+ \to \mathbb{R}$ is continuous and $\alpha$ is a constant. If $\alpha \leq 2, c > 0$ is a constant, and $r(t) \leq ct^{2-\alpha}$, then equation (1.22) is of the limit–point type. If $\alpha > 2$ and $r(t) \leq (2\alpha - 3)t^{\alpha-2}/4$, then equation (1.22) is of the limit–point type. Moreover, the constant $(2\alpha - 3)/4$ is sharp.

As a means of comparing some of the limit–circle and limit–point criteria described above, we present the following example.

**Example 1.2.** Consider the equation

$$(t^\alpha y')' + t^\beta y = 0, \quad t \geq 1, \tag{1.23}$$

where $\alpha$ and $\beta$ are constants. If we apply some of the results described above to this equation, we obtain the following:

   (i) Condition (1.9) of Dunford and Schwartz becomes $\alpha - \beta < 2$;

  (ii) Conditions (1.10) and (1.11) of Dunford and Schwartz become $\alpha + \beta > 2$ and $\alpha + \beta \leq 2$, respectively;

 (iii) Condition (1.12) of Everitt becomes $3\alpha - \beta < 6$;

 (iv) Condition (1.13) of Wong becomes $\alpha - \beta < 2$ which is identical to that of Dunford and Schwartz;

  (v) Condition (1.16) of Burton and Patula and condition (1.18) of Ráb become $\alpha = 0$ and $\beta > 2$;

 (vi) The limit–point criteria of Hartman and Wintner for $\alpha = 0$ becomes $\beta \leq 2$;

(vii) The limit–point criteria of Levinson becomes $\alpha + \beta \leq 2$ and $\alpha - \beta \leq 2$;

(viii) The limit–point criteria of Wong and Zettl becomes either (i) $2\alpha > 3$ and $\alpha - \beta > 2$ or (ii) $\alpha \geq 7/2$ and $\alpha - \beta = 2$. Notice how this agrees with the results in Example 1.1 above.

**Remark 1.3.** An extensive survey of results obtained prior to 1972 on the linear limit–point/limit–circle classification problem can be found in Knowles [81].

We close this section with a necessary and sufficient condition for equation (1.14) with $a \equiv 1$ to be of the limit–circle type that is due to Neuman [95, 96].

Let $v$ and $w$ be two linearly independent solutions of (1.7) with $a \equiv 1$. Then the continuous function $\alpha : \mathbb{R}_+ \to \mathbb{R}$ is called a *phase* of the basis $v, w$ if $\tan \alpha(t) = v(t)/w(t)$ for all $t \in \mathbb{R}_+$ satisfies $w(t) \neq 0$. Note (see [22, §5.5]) that $\alpha$ exists, $\alpha'(t) \neq 0$ on $\mathbb{R}_+$, and $\alpha \in C^{(3)}(\mathbb{R}_+)$. Moreover (see [22, §5.7]), $\alpha : \mathbb{R}_+ \to \mathbb{R}$ is a phase if and only if $\alpha \in C^3(\mathbb{R}_+)$, $\alpha'(t) \neq 0$ on $\mathbb{R}_+$, and

$$\frac{\alpha'''}{2\alpha'} - \frac{3}{4}\frac{\alpha''^2}{\alpha'^2} + \alpha'^2 - r(t) = 0, \quad t \in \mathbb{R}_+. \tag{1.24}$$

**Theorem 1.12.** ([96, Theorem 9.2.2]) *Equation* (1.14) *is of the limit–circle type if and only if for any phase* $\alpha$,

$$\int_0^\infty \frac{du}{|\alpha'(u)|} < \infty . \tag{1.25}$$

Using the above theorem, it is possible to construct examples of equations with prescribed asymptotic properties; for example, see Example 5.2 in Chapter 5 below.

# Chapter 2

# Basic Definitions

This chapter gives a description of the general nonlinear limit–point/limit–circle problem. A complication occurs because solutions may not be continuable to infinity, i.e., singular solutions may exist. These problems are discussed in this chapter as well.

## 2.1. Description of the Limit–Point/Limit–Circle Problem

Consider the $n$-th order nonlinear differential equation

$$y^{[n]} = r(t) f(y^{[0]}, y^{[1]}, \ldots, y^{[n-1]}), \tag{2.1}$$

where $n \geq 2$, $y^{[i]}$ is the $i$-th *quasiderivative* of $y$ defined as

$$y^{[0]} = \frac{1}{a_0(t)} y, \quad y^{[i]} = \frac{1}{a_i(t)} (y^{[i-1]})', \quad i = 1, \ldots, n-1, \quad y^{[n]} = (y^{[n-1]})',$$

the functions $a_i : \mathbb{R}_+ \to \mathbb{R}$, $i = 0, 1, \ldots, n-1$, and $f : \mathbb{R}^n \to \mathbb{R}$ are continuous,

$$r \in L_{\text{loc}}(\mathbb{R}_+), \quad a_i > 0, \quad \text{and} \quad x_1 f(x_1, \ldots, x_n) \geq 0 \quad \text{on} \quad \mathbb{R}^n. \tag{2.2}$$

Due to the way the functions $a_i$ often enter various integral conditions, it is convenient to write the coefficients in equations with quasiderivatives as reciprocals as we have done here. Together with (2.1), we consider $n$-th order nonlinear equations of the form

$$\mathcal{L}y \equiv \sum_{i=1}^{n} (-1)^i p_i(t) y^{(i)} = r(t) f(y, y', \ldots, y^{(n-1)}), \tag{2.3}$$

where $p_i$ are continuous on $\mathbb{R}_+$ and $p_n \neq 0$ on $\mathbb{R}_+$. If the coefficients $a_i$ in (2.1) are sufficiently smooth, then (2.1) can be written as (2.3). Conversely, if the equation $\mathcal{L}y = 0$ is nonoscillatory, then (2.3) can be written as (2.1).

A function $y$ defined on $J = [t_y, b) \subset \mathbb{R}_+$ is said to be a *solution* of equation (2.1) (of equation (2.3)) if $y^{[n-1]}$ (respectively $y^{(n-1)}$) is absolutely continuous on $J$, $y$ satisfies equation (2.1) (equation (2.3)) for almost all $t \in J$, and either $b = \infty$ or $b < \infty$ and $\sup_{t \in J} |y^{[n-1]}(t)| = \infty$ ($\sup_{t \in J} |y^{(n-1)}(t)| = \infty$). A solution $y$ is said to be *oscillatory* if it has an increasing sequence of zeros tending to $b$; otherwise, it is called *nonoscillatory*. A solution $y$ is called *continuable* if $J = \mathbb{R}_+$, and it is called *singular* if $b < \infty$. A continuable solution that is nontrivial in any neighborhood of $\infty$ is said to be a *proper* solution. Denote by $[\![a]\!]$ the greatest integer part of the number $a$.

**Remark 2.1.** (i) In this book, if an expression contains the $n$-th quasiderivative (the $n$-th derivative) of a solution, it is assumed that it holds for almost all $t$.

(ii) Note that a solution $y$ that is trivial in a neighborhood of $\infty$ is oscillatory according to our definition above.

Our main interest in this section consists in the investigation of continuable solutions only; for a discussion of continuability results see §2.2.

**Definition 2.1.** A continuable solution $y$ of equation (2.1) is said to be of the *nonlinear limit–circle type* if it satisfies

$$\int_0^\infty y^{[0]}(t) \, f(y^{[0]}(t), y^{[1]}(t), \ldots, y^{[n-1]}) \, dt < \infty. \tag{2.4}$$

Otherwise, it is of the *nonlinear limit–point type*, i.e.,

$$\int_0^\infty y^{[0]}(t) \, f(y^{[0]}(t), y^{[1]}(t), \ldots, y^{[n-1]}(t)) \, dt = \infty. \tag{2.5}$$

Equation (2.1) is said to be of the *nonlinear limit–circle type* if all its continuable solutions satisfy (2.4), and it is said to be of the *nonlinear limit–point type* if there exists a continuable solution $y$ for which (2.5) holds.

**Definition 2.2.** A continuable solution $y$ of equation (2.3) is said to be of the *nonlinear limit–circle type* if it satisfies

$$\int_0^\infty y(t) \, f(y(t), \ldots, y^{(n-1)}(t)) \, dt < \infty. \tag{2.6}$$

Otherwise, it is of the *nonlinear limit–point type*, i.e.,

$$\int_0^\infty y(t)\, f(y(t), \dots, y^{(n-1)}(t))\, dt = \infty. \tag{2.7}$$

Equation (2.3) is said to be of the *nonlinear limit–circle type* if all its continuable solutions satisfy (2.6), and it is said to be of the *nonlinear limit–point type* if there exists at least one continuable solution $y$ satisfying (2.7).

Note that Definitions 2.1 and 2.2 are not empty, that is, there exist equations satisfying each of these possibilities.

**Example 2.1.** The equation

$$y^{(n)} = y$$

is of the limit–point type since $y(t) = e^t$ is a solution of this equation. On the other hand, the equation

$$y^{(n)} = -y^3,$$

with $n$ either odd or of the form $4k$, $k = 1, 2, \dots$, is of the nonlinear limit–circle type (see [16, Theorem 3]).

**Remark 2.2.** Since the trivial solution $y \equiv 0$ is of the limit–circle type, the definition of equation (2.1) or equation (2.3) being of the limit–circle type does not depend on the existence of such a solution as it does in the case of limit–point type solutions.

**Remark 2.3.** In the case $\text{Im}\,\lambda \neq 0$, so that we are considering the linear equations (1.1) or (1.5), then a careful distinction must be made between the second and higher order cases. For second order linear equations, there is always one $L^2$ solution, so an equation being limit–circle means that there are two linearly independent $L^2$ solutions. As a consequence, second order linear equations are always either limit–circle or limit–point (see the discussion of Weyl's Alternative in Chapter 1). However, in the case of self-adjoint higher (even) order linear equations, say of order $2k$, if all solutions belong to $L^2$, we say the equation is (linear) limit–circle. There will always be $k$ linearly independent solutions belonging to $L^2$, and if there are only $k$, then the equation is said to be limit–point (see Section 1.3). Hence, in the case of higher even order linear equations, being not limit–circle means there is at least one solution not in $L^2$; this is not equivalent to the equation being limit–point.

Observe that if $\text{Im}\,\lambda = 0$ so that we are considering the linear equation (1.7) (or (1.6)), then we are no longer guaranteeing that there is one (or $n$) linearly independent solutions that belong to $L^2$.

In the remainder of this section, we will explain our choice of the form of the nonlinear limit–point/limit–circle terminology used above.

First, we consider the nonlinear equation

$$(a(t) y')' = r(t) f(y) \tag{2.8}$$

which is a special case of (2.1) with $n = 2$, $a_0 \equiv 1$, and $a_1 = \frac{1}{a}$, and we will assume that $r \leq 0$ on $\mathbb{R}_+$. This equation has been studied by many authors, and in Chapter 3 we present some of our results on this problem.

Special cases of (2.8), namely, the Emden–Fowler equations

$$y'' = -t^\sigma y^\gamma \tag{2.9}$$

where $\gamma \neq 1$ is the ratio of odd positive integers, and

$$y'' = r(t) y^{2n-1}, \tag{2.10}$$

$n = 2, 3, \ldots$, will serve as our motivating prototypes.

The exact form that integrability results for the nonlinear equation (2.8) should take is not completely obvious. While in the case of equation (2.10), showing that solutions belong to $L^{2n}$ seems to be the appropriate result, for equation (2.8) both

$$\int_0^\infty y(t) f(y(t)) \, dt < \infty \tag{2.11}$$

and

$$\int_0^\infty F(y(t)) \, dt < \infty, \tag{2.12}$$

where

$$F(y) = \int_0^y f(u) \, du, \tag{2.13}$$

agree with this choice and also agree with the square integrability of solutions of equation (1.7) as well.

**Remark 2.4.** If $f$ is nondecreasing, then

$$F(x) = \int_0^x f(u) du \leq f(x) \int_0^x du = x f(x),$$

and so condition (2.11) would then imply (2.12). As the following example shows, for more general classes of equations, conditions (2.11) and (2.12) are not equivalent.

**Example 2.2.** The function $y(t) = (t + 1)^{\frac{1}{2}}$, $t \geq 0$, is a solution of the equation

$$y'' = -\frac{1}{4}(t + 1)^{\frac{1}{2}} f(y)$$

where

$$f(x) = \begin{cases} \frac{\text{sgn}\, x}{x^4}, & \text{for } |x| \geq 1, \\ x, & \text{for } |x| < 1. \end{cases}$$

We see that

$$\int_0^\infty y(t)\, f(y(t))\, dt < \infty \quad \text{while} \quad \int_0^\infty \int_0^{y(t)} f(u)\, du\, dt = \infty.$$

The motivation for the first choice (2.11) is as follows. First, we restrict our attention to a special case of equation (2.1) with $a_i \equiv 1$, $i = 0, 1, \dots, n - 1$, namely,

$$y^{(n)} = r(t)\, f(y, y', \dots, y^{(n-1)}) \tag{2.14}$$

where $\alpha \in \{0, 1\}$, $r \in L_{\text{loc}}(\mathbb{R}_+)$, $f \in C^0(\mathbb{R}^n)$,

$$(-1)^\alpha r(t) \geq 0 \quad \text{on} \quad \mathbb{R}_+, \quad \text{and} \quad f(x_1, \dots, x_n)\, x_1 \geq 0 \quad \text{on} \quad \mathbb{R}^n. \tag{2.15}$$

Let $y$ be a continuable solution of equation (2.14). Define $n_0 = [\![\frac{n}{2}]\!]$ and

$$E(t) = \sum_{i=0}^{n_0-1} (-1)^{\alpha+i} y^{(n-i-1)}(t) y^{(i)}(t) + \frac{1}{2}(-1)^{\alpha+n_0}(n - 2n_0)(y^{(n_0)}(t))^2$$

$$+ (n - 2n_0 - 1)(-1)^{\alpha+n_0-1} \int_0^t (y^{(n_0)}(s))^2\, ds$$

for $t \in \mathbb{R}_+$. Then, in view of (2.15), it is easy to see that (see [2, Lemma 1.1])

$$E'(t) = (-1)^\alpha r(t)\, f(y(t), \dots, y^{(n-1)}(t))\, y(t) \geq 0 \quad \text{a.e. on} \quad \mathbb{R}_+.$$

Thus, $E$ is nondecreasing, and (2.15) yields

$$\int_0^\infty |r(s)|\, y(s)\, f(y(s), \dots, y^{(n-1)}(s))\, ds < \infty$$

if and only if $E$ is bounded on $\mathbb{R}_+$. \hspace{1cm} (2.16)

Hence, we can see that for $n = 2$, an integral similar to (2.11) with the weight function $|r|$ has been obtained.

An integral condition similar to (2.12) can be obtained for a special case of equation (2.14), namely,

$$y^{(n)} = r(t) f(y),\tag{2.17}$$

where $\alpha \in \{0, 1\}$, $r \in C^1(\mathbb{R}_+)$, $f \in C^0(\mathbb{R})$, $f(x) x \geq 0$ for $x \in \mathbb{R}$, $(-1)^\alpha r(t) \geq 0$ on $\mathbb{R}_+$, and either

$$r' \geq 0 \quad \text{or} \quad r' \leq 0 \quad \text{on} \quad \mathbb{R}.\tag{2.18}$$

To see this, define

$$
E_1(t) = (-1)^\alpha r(t) F(y(t)) + \sum_{i=0}^{n_0-1} (-1)^{i+\alpha+1} y^{(n-i-1)}(t) y^{(i+1)}(t)
$$

$$
+ \frac{1}{2}(-1)^{\alpha+n_0}(n-1-2n_0)(y^{(n_0)}(t))^2 + (n-2n_0)(-1)^{\alpha+n_0-1}
$$

$$
\times \int_0^t [y^{(n_0+1)}(s)]^2 \, ds, \quad t \in \mathbb{R}_+,
$$

where $F$ is given by (2.13). Then, a direct computation yields (see Lemma 3.11 in [2] for special values of $n$)

$$E_1'(t) = (-1)^\alpha r'(t) F(y(t)).$$

Thus, (2.18) implies $E_1$ is monotonic, and

$$\int_0^\infty |r'(t)| F(y(t)) \, dt < \infty \quad \text{if and only if} \quad E_1 \text{ is bounded on } \mathbb{R}_+.\tag{2.19}$$

In this case, a weighted integral of the form (2.12) is obtained.

We prefer investigating relation (2.11) rather than (2.12) because (2.16) holds for a larger class of equations than is covered by (2.19). Hence, we choose to define limit–point/limit–circle solutions and equations as in Definitions 2.1 and 2.1 above.

## 2.2. Continuable and Singular Solutions

In Definitions 2.1 and 2.2, we restrict our attention to continuable solutions only. However, equations (2.1) and (2.3) may have singular solutions as the following examples illustrate.

**Example 2.3.** The equation

$$y^{(4)} = 840 y^2 \, \text{sgn} \, y$$

has the singular solution $y(t) = (1 - t)^{-4}$ defined on $[0, 1)$.

**Example 2.4.** The equation

$$y'' = 2y^3$$

has the continuable solution $y(t) = (t+1)^{-1}$ and the singular solution $y = (1-t)^{-1}$ defined on $[0, 1)$.

Thus, we discuss the existence of continuable nontrivial solutions and of singular solutions. Moreover, the set of noncontinuable solutions ought to be nonempty if limit–point type equations are studied (see Remark 2.3 above). Hence, in the sequel, the following problems are investigated:

- the nonexistence of singular solutions, i.e., when all solutions are continuable and Definitions 2.1 and 2.2 apply to all solutions;

- the existence of at least one nontrivial continuable solution.

Furthermore, we will summarize results that are used later. The equations in Examples 2.3 and 2.4 are superlinear. Hence, it is natural to ask if, in the sublinear case, do singular solutions exit or are all solutions continuable. The following result gives easily verifiable conditions under which singular solutions of (2.1) and (2.3) do not exist.

**Theorem 2.1.** *Suppose there is a continuous nondecreasing function* $\omega : \mathbb{R}_+ \to \mathbb{R}_+$ *such that* $\int_1^\infty \frac{du}{\omega(u)} = \infty$ *and*

$$|f(x_1, \ldots, x_n)| \le \omega\left(\sum_{i=1}^n |x_i|\right) \quad on \quad \mathbb{R}^n. \tag{2.20}$$

*Then equations (2.1) and (2.3) do not have any singular solutions and so every solution is continuable.*

**Proof.** Equation (2.1) can be transformed into the system

$$y_i' = a_i(t)y_{i+1}, \quad i = 1, 2, \ldots, n-1,$$

$$y_n' = r(t) f(y_1, \ldots, y_n),$$

where $y_i = y^{[i-1]}$, $i = 0, 1, \ldots, n-1$. If the norm of $Y = (y_1, \ldots, y_n)^T$ is defined as $|Y| = \sum_{i=1}^n |y_i|$, then the statement follows from (2.20) and [72, Chapter 3, Theorem 5.1]. For equation (2.3), the proof is similar. $\qquad\square$

**Corollary 2.1.** *If equation* (2.1) *(equation* (2.3)*) is sublinear at infinity, i.e., if*

$$|f(x_1, \ldots, x_n)| \le \sum_{i=1}^{n} |x_i| \quad \text{for large } |x_i|, \ i = 1, \ldots, n, \tag{2.21}$$

*then every solution is continuable.*

Next, some sufficient conditions are stated under which continuable solutions of equation (2.1) are known to exist.

**Theorem 2.2.** ([27, Theorem 1]) *Let* $\alpha \in \{0, 1\}$, $n + \alpha$ *be even, and let*

$$(-1)^{\alpha} r(t) \ge 0 \quad \text{on} \quad \mathbb{R}_+. \tag{2.22}$$

*Then equation* (2.1) *has a nontrivial continuable solution y for which*

$$(-1)^i y^{[i]}(t) \, y(t) \ge 0 \quad \text{for} \quad t \in \mathbb{R}_+, \quad i = 1, 2, \ldots, n.$$

**Remark 2.5.** Consider equation (2.1) with $\alpha \in \{0, 1\}$, $n + \alpha$ odd, and (2.22) holding. The question of whether there exists at least one nontrivial continuable solution without additional assumptions placed on the equation has not been answered as yet even in the case $a_i \equiv 1, i = 1, 2, \ldots, n$.

A number of results concerning equation (2.14) with (2.15) holding are obtained in the monograph by Kiguradze and Chanturia [80]. We mention only one here. Again, we let $n_0 = [\![\frac{n}{2}]\!]$.

**Theorem 2.3.** *Let* $\alpha \in \{0, 1\}$, $n - n_0 - \alpha$ *be odd,* (2.15) *hold, and let continuous functions* $h : \mathbb{R}_+^{n_0} \to \mathbb{R}$ *and* $\omega : \mathbb{R}_+ \to \mathbb{R}$ *exist such that* $\omega$ *is nondecreasing,*

$$\int_1^{\infty} \frac{du}{\omega(u)} < \infty, \qquad \int_0^{\infty} t^{n-1} |r(t)| \, dt = \infty,$$

*and*

$$\omega(|x_1|) \le |f(x_1, \ldots, x_n)| \le h(|x_1|, \ldots, |x_{n_0}|) \quad \text{on} \quad \mathbb{R}^n.$$

*Then there exists a nontrivial continuable solution of equation* (2.14).

*Proof.* If equation (2.14) has a nontrivial continuable solution that is identically equal to zero in a neighborhood of $\infty$, then the statement holds. If such a solution does not exit, then the conclusion follows from Theorem 14.1 and Corollary 10.1 in [80]. $\square$

For second order differential equations, the nonexistence of singular solutions is ensured under weaker assumptions than in Theorem 2.1; see, for example, [63] or [80]. We give two theorems, one for the equation

$$(a(t)y')' = r(t) f(y) + e(t) \tag{2.23}$$

and one for the generalized Emden–Fowler equation

$$y'' = r(t)|y|^\lambda \operatorname{sgn} y, \quad \lambda > 0, \tag{2.24}$$

where $r$, $f$, and $e$ are continuous, $r(t) < 0$ on $\mathbb{R}_+$, and $xf(x) > 0$ for $x \neq 0$. The first result investigates the more general equation (2.23) if $r' \in L_{\text{loc}}(\mathbb{R}_+)$; in the second one, the existence of $r'$ is not assumed.

**Theorem 2.4.** ([63, Theorem 1]) *Let $a'$ exist and $r \in AC_{\text{loc}}(\mathbb{R}_+)$. Then every solution of (2.23) is continuable.*

**Theorem 2.5.** ([80, Theorem 17.1]) *Let $r$ be of bounded variation of any finite interval. Then every solution of (2.24) is continuable.*

Hence, in the case of $r < 0$, under very mild conditions on the function $r$, equation (2.23) has no singular solutions. In the opposite case, if $r \geq 0$, the situation is different. Equation (2.23) may have singular solutions even if $r$ is constant (see Example 2.4 above).

The following theorem gives us a sufficient condition under which singular solutions of equation (2.1) exist in case $r > 0$.

**Theorem 2.6.** ([28, Theorem 3] *Let $\lambda > 1$, $r(t) > 0$ on $\mathbb{R}_+$, and*

$$|x_1|^\lambda \leq |f(x_1, \ldots, x_n)| \quad on \ \mathbb{R}^n .$$

*Then there exists a nonoscillatory singular solution of equation (2.1).*

The following result gives sufficient conditions for the existence of oscillatory singular solutions for the case of equation (2.14) with a nonlinearity of the Emden–Fowler type, namely,

$$y^{(n)} = r(t)|y|^\lambda \operatorname{sgn} y . \tag{2.25}$$

**Theorem 2.7.** ([29, Theorem 1] *Let $\lambda > 1$, $r \in C^0(\mathbb{R}_+)$, and one of the following conditions hold:*

(i) $r(t) < 0$ *on* $\mathbb{R}_+$, *and either $n$ is odd or $n = 4k$, $k \in \{1, 2, \ldots\}$;*

(ii) $r(t) > 0$ *on* $\mathbb{R}_+$, $n \geq 5$, *and either $n$ or $n/2$ is odd.*

*Then there exists an oscillatory singular solution of (2.25).*

## 2.3. Extension of the Limit–Point/Limit–Circle Properties to Singular Solutions

Definitions 2.1 and 2.2 are restricted to continuable solutions only. Moreover, the problem of existence of limit–point or limit–circle type solutions depends only on their behavior in a neighborhood of $\infty$. Thus, a more general definition is needed to include all solutions. We describe the situation for equation (2.1); equation (2.3) can be studied similarly.

Since our purpose is to study the asymptotic behavior of solutions in a left-hand neighborhood of the right-hand endpoint of the interval of definition, we restrict our attention to solutions $y$ that are defined on $[t_y, T_y) \subset \mathbb{R}_+$. Note that according to the definition, $y$ can not be defined for $t \geq T_y$ if $T_y < \infty$.

**Definition 2.3.** A solution $y : [t_y, T_y) \to \mathbb{R}$ of equation (2.1) is said to be of the *nonlinear limit–circle type* if

$$\int_{t_y}^{T_y} y^{[0]}(t)\, f(y^{[0]}(t), \dots, y^{[n-1]}(t))\, dt < \infty ; \tag{2.26}$$

if

$$\int_{t_y}^{T_y} y^{[0]}(t)\, f(y^{[0]}(t), \dots, y^{[n-1]}(t))\, dt = \infty , \tag{2.27}$$

then $y$ is said to be of the *nonlinear limit–point type*. We will say that equation (2.1) is of the *nonlinear limit–circle type* if all its solutions satisfy (2.26), and it will be said to be of the *nonlinear limit–point type* if there is at least one solution $y$ satisfying (2.27).

It follows from Theorem 2.1 that Definition 2.1 and Definition 2.3 coincide if (2.20) holds. Furthermore, it is evident that if equation (2.1) is of the limit–circle type according to Definition 2.3, then it is of the limit–circle type according to Definition 2.1. That the converse relationship does not hold can be seen from the following results.

**Theorem 2.8.** *Let* (2.2) *hold,* $r \neq 0$ *be continuous on* $\mathbb{R}_+$, $a_0 \equiv 1$, *and let there exist a function* $g \in C^0(\mathbb{R}_+)$ *such that*

$$|f(x_1, \dots, x_n)| \leq g(|x_1|) \quad on \ \mathbb{R}^n . \tag{2.28}$$

*Then every singular solution $y$ of equation* (2.1) *is of the limit–point type in the sense of Definition 2.3.*

**Proof.** Let $y$ be a singular solution of (2.1) defined on $I = [t_y, b)$, $b < \infty$. Then, using (2.28),

$$J(t) \equiv \left| y^{[n-1]}(t) - y^{[n-1]}(t_y) \right| = \left| \int_{t_y}^{t} y^{[n]}(s)\, ds \right| \tag{2.29}$$

$$\leq \int_{t_y}^{t} |r(s)|\, |f(y^{[0]}, \ldots, y^{[n-1]}(s))|\, ds \leq M \int_{t_y}^{t} g\left( \left| \frac{y(s)}{a_0(s)} \right| \right) ds \,,$$

for $t \in I$ where $M = \max\limits_{t_y \leq s \leq b} |r(s)|$.

Since $y$ is a singular solution, $y^{[n-1]}$ is unbounded, and so $J$ is unbounded on $I$ as well. From this and from (2.29) we conclude that

$$y \quad \text{is unbounded on} \quad I \tag{2.30}$$

and

$$\int_{t_y}^{b} |f(y^{[0]}(s), \ldots, y^{[n-1]}(s))|\, ds = \infty. \tag{2.31}$$

Let $y$ be nonoscillatory. Then $y$ is monotone in a neighborhood of $b$ (see [2, Lemma 1.4]), and (2.30) yields the existence of $\tau \in I$ such that $|y(t)| \geq 1$ for $t \in [\tau, b)$. Thus, by (2.2) and (2.31),

$$\int_{t_y}^{b} y^{[0]}(t)\, f(y^{[0]}(t), \ldots, y^{[n-1]}(t))\, dt \geq \int_{\tau}^{b} |f(y^{[0]}(t), \ldots, y^{[n-1]}(t))|\, dt = \infty,$$

and so $y$ is of the limit–point type in the sense of Definition 2.3.

Now let $y$ be oscillatory. Set $N = \{t \in I : |\frac{y(t)}{a_0(t)}| \leq 1\}$ and observe that $N$ is the union of intervals (see [2, Theorem 2.4]). Then (2.2), (2.28), and (2.31) yield

$$\int_{t_y}^{b} y^{[0]}(t)\, f(y^{[0]}(t), \ldots, y^{[n-1]}(t))\, dt$$

$$\geq \int_{t_y}^{b} |f(y^{[0]}(t), \ldots, y^{[n-1]}(t))|\, dt - \int_{N} |f(y^{[0]}(t), \ldots, y^{[n-1]}(t))|\, dt$$

$$\geq \int_{t_y}^{b} |f(y^{[0]}(t), \ldots, y^{[n-1]}(t))|\, dt - \int_{N} g(|y(t)|)\, dt$$

$$\geq \int_{t_y}^{b} |f(y^{[0]}(t), \ldots, y^{[n-1]}(t))|\, dt - (b - t_y) \max_{|s| \leq 1} g(s) = \infty.$$

Thus, $y$ is of the limit–point type in the sense of Definition 2.3, and this completes the proof of the theorem. $\qquad\square$

The following example shows that singular solutions of the limit–circle type do exist.

**Example 2.5.** The equation

$$y^{(6)} = -\frac{1}{2}\left(\frac{945}{32}\right)^{-11} y(t)\left(y(t)^{(5)}\right)^{12} \operatorname{sgn} y(t)$$

has the singular solution $y(t) = (b - t)^{9/2}$, $b \in \mathbb{R}$, that is of the limit–circle type in the sense of Definition 2.3.

**Theorem 2.9.** *Let* $\lambda > 1$ *and* $n$ *be either odd or of the form* $n = 4k$ *for* $k \in \{1, 2, \dots\}$. *Suppose there exist positive constants* $K$ *and* $K_1$ *such that*

$$-K_1 \le r(t) \le -K < 0 \qquad for \qquad t \in \mathbb{R}_+.$$

*Then equation* (2.25) *is of the limit–circle type in the sense of Definition 2.1, and it is of the limit–point type in the sense of Definition 2.3.*

To prove this theorem, the following lemmas are needed. Let $L^\gamma([a, t])$ denote the space of Lebesque integrable functions $u$ defined on $[a, t]$ with $\|u\|_{L^\gamma} = \left(\int_a^t |u|^\gamma\right)^{1/\gamma}$.

**Lemma 2.1.** ([56, Theorem 1]) *Let* $\alpha, \beta, \gamma \ge 1$ *and let* $j \in \{0, 1, \dots, n - 1\}$ *be such that* $\frac{n-j}{\alpha} + \frac{j}{\beta} \ge \frac{n}{\gamma}$. *Then there exists* $\varrho > 0$ *such that*

$$\|u^{(j)}\|_{L^\gamma([a,t])} \le \varrho \left[x^{-j-\frac{1}{\alpha}+\frac{1}{\gamma}}\|u\|_{L^\alpha([a,t])} + x^{n-j-\frac{1}{\beta}+\frac{1}{\gamma}}\|u^{(n)}\|_{L^\beta([a,t])}\right]$$

*for arbitrary* $t > a$, $x \in (0, t - a]$, *and* $u \in C^n([a, t])$.

**Lemma 2.2.** *Let* $a \in \mathbb{R}_+$, $\lambda > 1$, *and let* $n$ *be either odd or of the form* $n = 4k$ *for* $k \in \{1, 2, \dots\}$. *Suppose there exist positive constants* $K$ *and* $K_1$ *such that*

$$-K_1 \le r(t) \le -K < 0 \quad for \quad t \in [a, a + 2].$$

*Then there exists* $M > 0$ *such that* $M$ *does not depend on* $a$ *and every solution* $y$ *of* (2.25) *satisfying the Cauchy initial conditions*

$$y^{(i)}(a) = C_i, \qquad i = 0, 1, \dots, n - 1, \tag{2.32}$$

*with*

$$\sum_{j=0}^{n_0-1}(-1)^{j-1}C_j\, C_{n-j-1} + (-1)^{n_0-1}\frac{n - 2n_0}{2}C_{n_0}^2 > M \tag{2.33}$$

*and*

$$\sum_{j=0}^{n-1}|C_j| > \sqrt{\frac{M}{n_0 + 1}}, \tag{2.34}$$

*where* $n_0 = [\![\frac{n}{2}]\!]$, *is singular and it is defined on* $I \subset [a, a + 2)$.

**Proof.** The statement of the lemma is in fact proved in the proof of Theorem 1 in [29] if we set $a = a$, $b = a+2$, $t_0 = a+1$, $\eta_1 = K$, and $\eta_2 = K_1$. Let $\varrho_j$ be given by Lemma 2.1 with $\alpha = \gamma = \lambda + 1$ and $\beta = \frac{\lambda+1}{\lambda}$ for each $j = 0, 1, 2, \ldots, n-1$, and let $\varrho = \max_{0 \le j \le n-1}\{\varrho_j\}$. Note that $\varrho$ does not depend on $a$ as can be seen from the transformation $\bar{t} = t + \text{const.}$ Set

$$v = 4\varrho^2(n - n_0)K_1^{(n-1)/(n-\frac{\lambda-1}{\lambda+1})}, \qquad \delta = \sum_{k=0}^{n-1}(k+1)!(2k+1)^{k+1},$$

and

$$\mu = ((\lambda + 1)n - \lambda + 1)\left((\lambda + 1)n - \lambda + 1 - 2\frac{\lambda - 1}{\lambda + 1}\right)^{-1} > 1.$$

In the proof of Theorem 1 in [29], it is shown that a solution $y$ of equation (2.25) satisfying (2.32) with

$$\sum_{j=0}^{n_0-1}(-1)^{j-1}C_jC_{n-j-1} + (-1)^{n_0-1}\frac{n-2n_0}{2}C_{n_0}^2 > K^{-\frac{1}{\mu-1}}v^{\frac{\mu}{\mu-1}}(\mu-1)^{-\frac{2}{(\lambda+1)(\mu-1)}}$$

and

$$\sum_{j=0}^{n-1}|C_j| > K^{-\frac{\lambda}{\lambda-1}}(\delta K + K_1)$$

is singular and it is defined on $I \subset [a, a + 2)$. From this it is easy to see that $M$, not depending on $a$, exists. $\qquad\square$

**Proof of Theorem 2.9.** According to Theorem 2.7, the set of singular solutions of equation (2.25) is nonempty, and Theorem 2.8 yields that every singular solution is of the limit–point type in the sense of Definition 2.3. Hence, equation (2.25) is of the limit–point type in the sense of Definition 2.3.

Now we prove that every continuable solution of equation (2.25) is of the limit–circle type. Thus, let $y$ be a continuable solution of (2.24), set $n_0 = [\![\frac{n}{2}]\!]$, and define

$$F(t) = \sum_{i=0}^{n_0-1}(-1)^{i-1}y^{(n-i-1)}(t)y^{(i)}(t) + (-1)^{n_0-1}\frac{n-2n_0}{2}[y^{n_0}(t)]^2 \qquad (2.35)$$

for $t \in \mathbb{R}_+$. Then,

$$F'(t) = -r(t)|y(t)|^{\lambda+1} - (n - 2n_0 - 1)[y^{(n_0)}(t)]^2 \ge 0 \qquad (2.36)$$

for $t \in \mathbb{R}_+$. Thus, $F$ is nondecreasing.

Since the hypotheses of Lemma 2.2 are satisfied, let $M$ denote the constant given by that lemma. We will show that

$$F(t) \leq M, \quad t \in \mathbb{R}_+ . \tag{2.37}$$

To the contrary, suppose there exists $t_0 \in \mathbb{R}_+$ such

$$F(t_0) > M . \tag{2.38}$$

From the fact that (2.33) holds for $a = t_0$, Lemma 2.2 implies (2.34) does not hold, i.e.,

$$\sum_{j=0}^{n-1} |y^{(j)}(t_0)| \leq \sqrt{\frac{M}{n_0 + 1}} .$$

But from this and from (2.35), we have

$$F(t_0) \leq \frac{M}{n_0 + 1}(n_0 + 1) = M .$$

This contradicts (2.38) and so (2.37) holds for $t \in \mathbb{R}_+$. Furthermore, (2.35) and (2.36) yield

$$\int_0^\infty |y(t)|^{\lambda+1} \, dt \leq \int_0^\infty \frac{F'(t) \, dt}{|r(t)|} \leq \frac{1}{K} \int_0^\infty F'(t) \, dt = \frac{1}{K}(F(\infty) - F(0))$$
$$\leq \frac{1}{K}(M - F(0)) < \infty,$$

and hence $y$ is of the limit–circle type. It follows that equation (2.24) is of the limit–circle type according to Definition 2.1. □

Theorem 2.9 shows that Definition 2.1 and Definition 2.3 are clearly different. Definition 2.1 is more frequently used since continuable solutions are in general more important in applications than are singular ones. Moreover, Theorem 2.8 shows that equation (2.1) is of the limit–point type in the sense of Definition 2.3 nearly always if singular solutions exist.

**Open Problems.**

**Problem 2.1.** *Do nonoscillatory singular solutions of equation* (2.25) *exist?*

**Problem 2.2.** *Do there exist singular solutions (oscillatory or nonoscillatory) of equation* (2.14) *with $n \geq 2$ and $r < 0$?*

**Problem 2.3.** *Determine sufficient conditions for the existence of nontrivial continuable solutions of equation* (2.1) *with* $\alpha \in \{0, 1\}$, $n + \alpha$ *odd, and* (2.22) *holding.*

**Notes.** Examples 2.1, 2.2, and 2.5 can be found in Bartušek, Došlá, and Graef [16]; Examples 2.3 and 2.4 are new. Theorem 2.1 is a generalization of a result due to Wintner and can be found in Hartman's book [72]. Theorems 2.8 and 2.9 and Lemma 2.2 are due to Bartušek, Došlá, and Graef [16].

# Chapter 3

# Second Order Nonlinear Equations

In this chapter, we introduce the limit–point/limit–circle problem for second order nonlinear equations and describe the current state of research on this problem. There are a number of papers in the literature that discuss the existence of an $L^2$ solution to a nonlinear equation, but our focus throughout this work is on results that imply that all solutions have this or an analogous property.

## 3.1. Introduction

Here, we consider the second order nonlinear equation

$$(a(t)y')' + r(t)f(y) = 0 \tag{3.1}$$

where $a, r : \mathbb{R}_+ \to \mathbb{R}$ and $f : \mathbb{R} \to \mathbb{R}$ are continuous, $a', r' \in AC_{loc}(\mathbb{R}_+)$, $a''$, $r'' \in L^2_{loc}(\mathbb{R}_+)$, $a(t) > 0$, $r(t) > 0$ and $yf(y) \geq 0$ for all $y$. These smoothness conditions imposed on the functions $a$ and $r$ are sufficient to guarantee the continuability of all solutions of equation (3.1) (see Theorem 2.4 above). For a further discussion of continuability results for equation (3.1) (as well as the boundedness and convergence to zero of solutions) under conditions compatible with those used here, we refer the reader to the papers of Graef et al. [61, 62, 63]. In view of this and our discussion in Chapter 2, we will then say that a solution $y(t)$ of (3.1) is of the *nonlinear limit–circle type* if

$$\int_0^\infty y(t) f(y(t)) \, dt < \infty \tag{3.2}$$

and it is of the *nonlinear limit–point type* otherwise, i.e.,

$$\int_0^\infty y(t)\, f(y(t))\, dt = \infty. \tag{3.3}$$

Equation (3.1) will be said to be of the *nonlinear limit–circle type* if all its solutions satisfy (3.2), and it will be said to be of the *nonlinear limit–point type* if there is at least one solution satisfying (3.3).

As indicated in Chapter 1, the study of the nonlinear counterpart of the (linear) limit–point/limit–circle problem is of recent origin. In fact, the only references in the nonlinear case appear to be the papers of Atkinson [1], Burlak [23], Detki [32], Elias [43], Graef [58, 59, 60], Graef and Spikes [64, 65, 66], Hallam [70], Kroopnick [88], Spikes [104, 105], Suyemoto and Waltman [107], and Wong [116]. Furthermore, with the exception of those of Graef and Spikes [58, 59, 60, 65, 66, 104, 105], all the results for nonlinear equations are of the limit–point type. In the next chapter, we will relate some of the conclusions in these papers to our results here.

The nonlinear limit–point/limit–circle results presented here can actually be obtained for equations with a more general nonlinear term, and in this regard we refer the reader to the papers of Graef [59, 60], Spikes [104, 105], and Graef and Spikes [65]. Moreover, many of the theorems in this chapter and Chapter 5 also hold for equations with a forcing term or a more general perturbation term; again see [59, 60, 65, 104, 105] as well as Sections 3.2.5 and 3.3.4 below. However, the added generality requires additional assumptions on the coefficient functions (that dissappear when reduced to the equations considered here), and this only tends to obscure the beauty of the results. Thus, in this chapter and in Chapter 5, we only consider Emden–Fowler type nonlinearities in equations (3.4) and (3.54) below.

## 3.2. The Superlinear Equation

To simplify the discussion which follows, we will restrict our attention to the equation

$$(a(t)y')' + r(t)y^{2k-1} = 0 \tag{3.4}$$

where $k > 1$ is a positive integer. Later, we will discuss the more general nonlinear equation (3.1) and also some other extensions.

### 3.2.1. Limit–Circle Criteria

Making a transformation of the independent variable $t$ has also proved to be useful in the study of the nonlinear equation (3.1). For example, applying (1.8) to (3.1) yields the equation

$$\ddot{x} + 2p(t)\dot{x} + f(x) = 0$$

where $p(t) = [a(t)r(t)]'/4a^{\frac{1}{2}}(t)r^{\frac{3}{2}}(t)$ as before. (See, for example, Graef and Spikes [65] and Spikes [105].) Other transformations have also been effectively used on equation (3.1) (see Graef [58, 59, 60]).

To simplify notation, we let $\alpha = 1/2(k+1)$ and $\beta = (2k+1)/2(k+1)$. We then make the transformation

$$s = \int_0^t [r^\alpha(u)/a^\beta(u)]du, \quad x(s) = y(t) \tag{3.5}$$

so that equation (3.4) becomes

$$\ddot{x} + \alpha p(t)\dot{x} + P(t)x^{2k-1} = 0 \tag{3.6}$$

where

$$p(t) = (a(t)r(t))'/a^\alpha(t)r^{\alpha+1}(t) \quad \text{and} \quad P(t) = (a(t)r(t))^{\beta-\alpha}.$$

Note also that

$$\beta - \alpha = 2\beta - 1 = k/(k+1).$$

The form of this transformation is motivated by the shape of the nonlinear term in equation (3.4). When $k = 1$, it does not reduce to the transformations used by other authors for equation (1.7). Equation (3.6) can then be written as the system

$$\dot{x} = z - \alpha p(t)x,$$
$$\dot{z} = \alpha \dot{p}(t)x - P(t)x^{2k-1}. \tag{3.7}$$

We are now ready to prove our first result for equation (3.4).

**Theorem 3.1.** *Assume that*

$$\int_0^\infty \left\{ \left| \{(a(u)r(u))'/a^\alpha(u)r^{\alpha+1}(u)\}' \right| /[a(u)r(u)]^{(\beta-\alpha)/2} \right\} du < \infty, \tag{3.8}$$

$$\int_0^\infty \left| \{(a(u)r(u))'/a^\alpha(u)r^{\alpha+1}(u)\}' \right| [a(u)r(u)]^{(\beta-\alpha)/2} du < \infty, \tag{3.9}$$

*and*

$$\int_0^\infty [1/(a(u)r(u))^{\beta-\alpha}]du < \infty. \tag{3.10}$$

*Then equation (3.4) is of the nonlinear limit–circle type, i.e., any solution $y(t)$ of (3.4) satisfies*

$$\int_0^\infty y^{2k}(u)du < \infty. \tag{3.11}$$

*Proof.* Define
$$V(x, z, s) = z^2/2 + P(t)x^{2k}/2k;$$
then
$$\dot{V} = \alpha \dot{p}(t)xz - P(t)x^{2k-1}z + P(t)x^{2k-1}[z - \alpha p(t)x] + \dot{P}(t)x^{2k}/2k$$
$$= \alpha \dot{p}(t)xz + [\dot{P}(t)/2k - \alpha p(t)P(t)]x^{2k}$$
$$= \alpha \dot{p}(t)xz. \tag{3.12}$$

Now
$$|xz| = |P^{\frac{1}{2}}(t)xz|/P^{\frac{1}{2}}(t)$$
$$\leq [P(t)x^2/2 + z^2/2]/P^{\frac{1}{2}}(t)$$
$$\leq [P(t)(x^{2k}/2k + K_1) + z^2/2]/P^{\frac{1}{2}}(t)$$
$$\leq V(s)/P^{\frac{1}{2}}(t) + K_1 P^{\frac{1}{2}}(t)$$

for some constant $K_1 \geq 0$. Thus, we have
$$\dot{V}(s) \leq \alpha |\dot{p}(t)|V(s)/P^{\frac{1}{2}}(t) + \alpha |\dot{p}(t)|K_1 P^{\frac{1}{2}}(t). \tag{3.13}$$

Since $\dot{p}(t) = p'(t)a^\beta(t)/r^\alpha(t)$, if we let $\tau(s)$ denote the inverse function of $s(t)$, we have
$$\int_0^s \left\{ |\dot{p}(\tau(v))|/P^{\frac{1}{2}}(\tau(v)) \right\} dv$$
$$= \int_0^t \left\{ |\{a(u)r(u))'/a^\alpha(u)r^{\alpha+1}(u)\}'|/[a(u)r(u)]^{(\beta-\alpha)/2} \right\} du.$$

Also,
$$\int_0^s |\dot{p}(\tau(v))| P^{\frac{1}{2}}(\tau(v))dv$$
$$= \int_0^t \left\{ |\{a(u)r(u))'a^\alpha(u)r^{\alpha+1}(u)\}'|[a(u)r(u)]^{(\beta-\alpha)/2} \right\} du.$$

Integrating (3.13), we obtain
$$V(s) \leq V(0) + \alpha \int_0^s \left\{ |\dot{p}(\tau(v))|/P^{\frac{1}{2}}(\tau(v)) \right\} V(v)dv$$
$$+ K_1\alpha \int_0^s |\dot{p}(\tau(v))| P^{\frac{1}{2}}(\tau(v))dv. \tag{3.14}$$

Condition (3.9) implies that the second integral on the right-hand side of (3.14) converges, so applying Gronwall's inequality we have

$$V(s) \le M_1 \exp \int_0^s \left\{ |\dot{p}(\tau(v))|/P^{\frac{1}{2}}(\tau(v)) \right\} dv,$$

for some constant $M_1 > 0$. By condition (3.8), the above integral converges, and we have that $V(s)$ is bounded, say, $V(s) \le M_2$ for some $M_2 > 0$. Therefore,

$$(a(t)r(t))^{\beta-\alpha} y^{2k}(t) = (a(t)r(t))^{\beta-\alpha} x^{2k}(s) \le 2kM_2,$$

from which it follows that

$$\int_0^\infty y^{2k}(u)du \le 2kM_2 \int_0^\infty [1/(a(u))r(u))^{\beta-\alpha}]du < \infty$$

by condition (3.10), and so all solutions of equation (3.4) are of the nonlinear limit–circle type.                                                                               $\square$

In order to discuss the content of Theorem 3.1, first let $a(t) \equiv 1$ so that (3.4) becomes

$$y'' + r(t)y^{2k-1} = 0. \tag{3.15}$$

**Corollary 3.1.** *Assume that*

$$\int_0^\infty |\{r'(u)/r^{\alpha+1}(u)\}'| /r^{(\beta-\alpha)/2}(u)\}du < \infty, \tag{3.16}$$

$$\int_0^\infty |\{r'(u)/r^{\alpha+1}(u)\}'| r^{(\beta-\alpha)/2}(u)\}du < \infty, \tag{3.17}$$

*and*

$$\int_0^\infty [1/r(u)]^{\beta-\alpha}]du < \infty. \tag{3.18}$$

*Then equation (3.15) is of the nonlinear limit–circle.*

In terms of $k$, the above conditions become

$$\int_0^\infty \left\{ |\{r'(u)/r^{\frac{2k+3}{2(k+1)}}(u)\}'| /r^{\frac{k}{2(k+1)}}(u) \right\} du < \infty, \tag{3.19}$$

$$\int_0^\infty |\{r'(u)/r^{\frac{2k+3}{2(k+1)}}(u)\}'| r^{\frac{k}{2(k+1)}}(u)\}du < \infty, \tag{3.20}$$

*and*

$$\int_0^\infty [1/r(u)]^{\frac{k}{k+1}}du < \infty. \tag{3.21}$$

The first two of these can be rewritten as

$$\int_0^\infty \left\{ \left| r''(u)/r^{\frac{3}{2}}(u) - \frac{2k+3}{2(k+1)} [r'(u)]^2/r^{\frac{5}{2}}(u) \right| \right\} du < \infty, \tag{3.22}$$

$$\int_0^\infty \left\{ \left| r''(u)/r^{\frac{k+3}{2(k+1)}}(u) - \frac{2k+3}{2(k+1)} [r'(u)]^2/r^{\frac{3k+5}{2(k+1)}}(u) \right| \right\} du < \infty. \tag{3.23}$$

It is important to note that if $k = 1$ (equation (3.4) is linear), then in the proof of Theorem 3.1, $K_1 = 0$, and so condition (3.9) is not needed. As a consequence, (3.17) ((3.20) and (3.23) as well) is not needed. Moreover, we see that (3.22) is exactly condition (1.15) above of Dunford and Schwartz, and $\beta - \alpha = k/(k+1) = 1/2$ (see (1.10)).

To see that condition (3.10) is sharp, consider the special case of (3.15) when $r(t) = t^\sigma$, namely,

$$y'' + t^\sigma y^{2k-1} = 0. \tag{3.24}$$

Now (3.10) (see (3.21)) implies that $\sigma k/(k+1) > 1$, or

$$\sigma > 1 + 1/k. \tag{3.25}$$

From asymptotic integrations of equation (3.24), it is known that all solutions of (3.24) belong to $L^{2k}(\mathbb{R}_+)$ only if (3.25) holds (see Atkinson [1, p. 311] or Bellman [21, p. 163]). Since (3.22) and (3.23) are obviously satisfied in this case, we have that (3.25) is both necessary and sufficient for all solutions of (3.24) to belong to $L^{2k}(\mathbb{R}_+)$, i.e., to be of the nonlinear limit–circle type.

### 3.2.2. Necessary Conditions for Limit–Circle Behavior

The next lemma gives a necessary condition for a solution of (3.4) to be of the nonlinear limit–circle type. We will use it to obtain a limit–point result.

**Lemma 3.1.** *Assume that there exists $N_2 > 0$ such that*

$$\left| (a(t)r(t))'/a^{1/2}(t)r^{3/2}(t) \right| \leq N_2 \tag{3.26}$$

*and*

$$\int_0^\infty \left\{ \left[ (a(u)r(u))' \right]^2 /a(u)r^3(u) \right\} du < \infty. \tag{3.27}$$

*If $y(t)$ is a nonlinear limit–circle type solution of equation (3.4), then*

$$\int_0^\infty \{ [(a(u)r(u))']^2 y^2(u)/a(u)r^3(u) \} du < \infty.$$

*Proof.* We have

$$\int_0^\infty \{[(a(u)r(u))']^2 y^2(u)/a(u)r^3(u)\}du$$

$$\le N_2^2 \int_0^\infty [y^{2k}(u)]du + \int_0^\infty \{[(a(u)r(u))']^2/a(u)r^3(u)\}du < \infty$$

by (3.11) and (3.27). $\qquad\square$

**Theorem 3.2.** *Suppose that there exists a constant $N_1 > 0$ such that*

$$|a^{\frac{1}{2}}(t)r'(t)/r^{\frac{3}{2}}(t)| \le N_1 \tag{3.28}$$

*and*

$$\int_0^\infty \{a(u)[r'(u)]^2/r^3(u)\}du < \infty. \tag{3.29}$$

*If $y(t)$ is a nonlinear limit–circle type solution of equation (3.4), i.e.,*

$$\int_0^\infty y^{2k}(u)du < \infty, \tag{3.30}$$

*then*

$$\int_0^\infty \{a(u)[y'(u)]^2/r(u)\}du < \infty. \tag{3.31}$$

*Proof.* If we multiply equation (3.4) by $y(t)/r(t)$, use the identity $(a(t)y')'y = (a(t)y'y)' - a(t)[y']^2$, and integrate by parts, we obtain

$$a(t)y'(t)y(t)/r(t) - a(t_1)y'(t_1)y(t_1)/r(t_1)$$

$$+ \int_{t_1}^t [a(u)y'(u)y(u)r'(u)/r^2(u)]du + \int_{t_1}^t y^{2k}(u)du$$

$$- \int_{t_1}^t \{a(u)[y'(u)]^2/r(u)\}du = 0 \tag{3.32}$$

for any $t_1 \ge 0$. By the Schwarz inequality,

$$\left| \int_{t_1}^t [a(u)y'(u)y(u)r'(u)/r^2(u)]du \right|$$

$$\le \left[ \int_{t_1}^t \{a(u)[y'(u)]^2/r(u)\}du \right]^{\frac{1}{2}} \left[ \int_{t_1}^t \{a(u)y^2(u)/[r'(u)]^2/r^3(u)\}du \right]^{\frac{1}{2}}.$$

Now from (3.28) we have

$$a(t)y^2(t)[r'(t)]^2/r^3(t) \le \{a(t)[r'(t)]^2/r^3(t)\}[y^{2k}(t) + 1]$$
$$\le N_1^2 y^{2k}(t) + a(t)[r'(t)]^2/r^3(t),$$

so integrating and applying (3.29) and (3.30), we obtain

$$\int_{t_1}^{\infty} \{a(u)y^2(u)[r'(u)]^2/r^3(u)\}du \le K_2 < \infty.$$

If $y(t)$ is not eventually monotonic, let $\{t_j\} \to \infty$ be an increasing sequence of zeros of $y'(t)$. Then from (3.32) we have

$$K_2 H^{\frac{1}{2}}(t_j) + K_3 \ge H(t_j)$$

where

$$H(t) = \int_{t_1}^{t} \{a(u)[y'(u)]^2/r(u)\}du$$

and $K_3 > 0$ is a constant. It follows that $H(t_j) \le K_4 < \infty$ for all $j$ and some constant $K_4 > 0$, so (3.31) holds.

If $y(t)$ is eventually monotonic, then $y(t)y'(t) \le 0$ for $t \ge t_1$ for sufficiently large $t_1 \ge 0$ since otherwise condition (3.30) would be violated. Using this fact in (3.32) we can repeat the type of argument used above to again obtain that (3.31) holds. $\qquad\square$

**Remark 3.1.** It is not surprising that conditions like (3.26) and (3.28) arise here since similar conditions occur in the asymptotic analysis of solutions of second order linear equations. For example, Fedoryuk [54] studies the linear equation (1.14) with $a(t) \equiv 1$ and requires that

$$\lim_{t \to \infty} \frac{r'(t)}{r^{\frac{3}{2}}(t)}$$

exists as a finite number (compare this to (3.26) or (3.28)). Fedoryuk [54] also uses the condition

$$\int_0^{\infty} \left( \frac{r'(u)}{r^{3/2}(u)} \right)' du < \infty$$

which is similar to condition (1.15) of Dunford and Schwartz. Burton and Patula [24, Lemma 2] show that if $a(t) \equiv 1$ and (3.28) holds, then (3.31) is a necessary condition for $y$ to be a limit–circle type solution of the linear equation (1.14). For the case $a(t) \equiv 1$, also see the Lemma in Section 6 of Hartman and Wintner [73].

### 3.2.3. Limit–Point Criteria

Limit–point criteria for second order linear equations can be found in the works of many authors. Among the best known results of this type are those of Hartman and Wintner [73, 74], Levinson [92], and Wintner [115] (see Chapter 4 below as well as the discussion in Section 1.3 above). Titchmarsh [109, 110] discusses an important relationship between the existence of a limit–point solution and the solution of certain boundary value problems. The importance of obtaining that equation (3.4) is of limit–point type is pointed out in [73]. As a consequence of one of the limit–point theorems given in this section, we are able to obtain necessary and sufficient conditions for a special case of equation (3.4) to be limit–circle.

First, we introduce the following notation. For any continuous function $h$ we let

$$h(u)_+ = \max\{h(u), 0\} \quad \text{and} \quad h(u)_- = \max\{-h(u), 0\}$$

so that

$$h(u) = h(u)_+ - h(u)_-.$$

**Theorem 3.3.** *Suppose that* (3.26)–(3.27) *and either*

$$\int_0^\infty [1/a(u)r(u)] \exp\left(-\int_0^u [(a(v)r(v))'_-/a(v)r(v)]dv\right) du = \infty \qquad (3.33)$$

*or*

$$\int_0^\infty \exp\left(-\int_0^u [(a(v)r(v))'_+/a(v)r(v)]dv\right) du = \infty \qquad (3.34)$$

*hold. Then equation* (3.4) *is of the nonlinear limit–point type, i.e., there is a solution of* (3.4) *that does not satisfy* (3.11).

*Proof.* We write equation (3.4) as the system

$$y' = w, \quad w' = (-a'(t)w - r(t)y^{2k-1})/a(t). \qquad (3.35)$$

If (3.33) holds, define

$$V(y, w, t) = a^2(t)w^2/2 + a(t)r(t)y^{2k}/2k.$$

Then,

$$V'(t) = (a(t)r(t))'y^{2k}/2k \geq -[(a(t)r(t))'_-/a(t)r(t)]V. \qquad (3.36)$$

Let $(y(t), w(t))$ be a solution of (3.35) with $(y(0), w(0)) = (y_0, w_0)$ and $V(y_0, w_0, 0) = V(0) > 1$. Then from (3.36), we have

$$\left(V(t) \exp \int_0^t [(a(u)r(u))'_-/a(u)r(u)]du\right)' \geq 0.$$

Integrating, we obtain

$$V(t) \exp \int_0^t [(a(u)r(u))'_-/a(u)r(u)]du \geq V(0) > 1.$$

Hence,

$$\int_0^t [V(u)/a(u)r(u)]du$$

$$\geq \int_0^t [1/a(u)r(u)] \exp\left(-\int_0^u [(a(v)r(v))'_-/a(v)r(v)dv\right)du \to \infty$$

as $t \to \infty$. In view of Theorem 3.2 this shows that $y(t)$ cannot be a limit–circle solution of (3.4).

If (3.34) holds, define $V_1(y, w, t) = a(t)w^2/2r(t) + y^{2k}/2k$. Then,

$$V_1' \geq [(a(t)r(t))_+/a(t)r(t)]V_1.$$

The remainder of the proof is similar to the proof when (3.33) holds and is omitted.                                                                                     □

**Remark 3.2.** Wong and Zettl [121, Theorem 1] proved a similar result for linear equations under conditions which imply (3.33) and (3.34) of Theorem 3.3, but, on the other hand, they did not require condition (3.26).

**Theorem 3.4.** *Suppose conditions* (3.8), (3.9), *and* (3.26)–(3.29) *hold. If*

$$\int_0^\infty [1/(a(u)r(u))^{\beta-\alpha}]du = \infty, \tag{3.37}$$

*then equation* (3.4) *is of the nonlinear limit–point type.*

*Proof.* As in the proof of Theorem 3.1, define

$$V(x, z, s) = z^2/2 + (a(t)r(t))^{\beta-\alpha}x^{2k}/2k$$

and differentiate to obtain

$$\dot{V}(s) \geq -\alpha|\dot{p}(t)|V(s)/P^{\frac{1}{2}}(t) - \alpha|\dot{p}(t)|K_1 P^{\frac{1}{2}}(t).$$

We then have

$$\dot{V}(s) + \alpha[|\dot{p}(t)|/P^{\frac{1}{2}}(t)]V(s) \geq -\alpha|\dot{p}(t)|K_1 P^{\frac{1}{2}}(t).$$

If we define the functions $H$ and $h : \mathbb{R}_+ \to \mathbb{R}$ by

$$H(t) = \alpha |\dot{p}(t)|/P^{\frac{1}{2}}(t)$$

and

$$h(t) = \alpha |\dot{p}(t)| K_1 P^{\frac{1}{2}}(t),$$

we can then write

$$\dot{V}(s) + H(t)V(s) \geq -h(t).$$

Hence,

$$\frac{d}{ds}\left(V(s)\exp\int_0^s H(\tau(\xi))d\xi\right) \geq -h(t)\exp\int_0^s H(\tau(\xi))d\xi. \tag{3.38}$$

Now condition (3.8) guarantees that

$$\exp\int_0^\infty H(\tau(\xi))d\xi \leq K_6 < \infty$$

for some constant $K_6 > 0$, while condition (3.9) implies that

$$K_6\int_0^\infty h(\tau(\xi))d\xi \leq K_7 < \infty$$

for some $K_7 > 0$.

Let $y(t)$ be any solution of (3.4) such that $V(x(0), z(0), 0) > K_7 + 1$. Integrating (3.38), we have

$$V(s)\exp\int_0^s H(\tau(\xi))d\xi \geq V(0) - K_7 > 1,$$

and so

$$V(s) \geq 1/K_6$$

for $s \geq 0$. Dividing both members of this last inequality by $(a(t)r(t))^{\beta-\alpha}$ and rewriting the left-hand side in terms of $t$, we have

$$a(t)[y'(t)]^2/2r(t) + \alpha(a(t)r(t))'y(t)y'(t)/r^2(t)$$
$$+ \alpha^2[(a(t)r(t))']^2y^2(t)/2a(t)r^3(t) + y^{2k}/2k \geq 1/K_6(a(t)r(t))^{\beta-\alpha}. \tag{3.39}$$

If $y(t)$ is a limit–circle solution of (3.4), then Theorem 3.2 implies

$$\int_0^\infty \{a(u)[y'(u)]^2/r(u)\}du < \infty,$$

and Lemma 3.1 implies

$$\int_0^\infty \{[(a(u)r(u))']^2 y^2(u)/a(u)r^3(u)\}du < \infty.$$

By the Schwarz inequality,

$$\left|\int_0^\infty \{(a(u)r(u))'y(u)y'(u)/r^2(u)\}\,du\right|$$

$$\leq \left[\int_0^\infty \{[a(u)r(u))']^2 y^2(u)/a(u)r^3(u)\}\,du\right]^{\frac{1}{2}}$$

$$\times \left[\int_0^\infty \{a(u)[y'(u)]^2/r(u)\}\,du\right]^{\frac{1}{2}} < \infty$$

again by Lemma 3.1 and Theorem 3.2. Finally, since $y(t)$ is a limit–point type solution, an integration of (3.39) yields a contradiction to (3.37).  □

**Remark 3.3.** If $r(t) = t^\sigma$ in equation (3.15) so that we actually have the equation (3.24),

$$y'' + t^\sigma y^{2k-1} = 0,$$

then

$$\int_0^\infty [1/r^{k/(k+1)}(u)]du = \infty$$

implies that $\sigma k/(k+1) \leq 1$, or,

$$\sigma \leq 1 + 1/k$$

(compare this with (3.25)). This is in complete agreement with Atkinson's results [1], and leads us to believe that condition (3.37) is sharp.

**Remark 3.4.** In view of condition (1.19) in Levinson's theorem on linear equations, Theorem 1.10 above, it is not surprising that conditions like (3.26) and (3.28) appear in a result guaranteeing that equation (3.4) be of the nonlinear limit–point type.

### 3.2.4. Necessary and Sufficient Conditions

By combining Theorem 3.1 with Theorem 3.4 we can obtain necessary and sufficient conditions for equation (3.4) to be of the nonlinear limit–circle type.

**Theorem 3.5.** *Assume that conditions* (3.8)–(3.9) *and* (3.26)–(3.29) *hold. Then equation* (3.4) *is of the nonlinear limit–circle type if and only if*

$$\int_0^\infty [1/(a(u))r(u))^{k/(k+1)}]du < \infty. \tag{3.40}$$

When we specialize this theorem to equation (3.15), we obtain the following result.

**Corollary 3.2.** *Assume that*

$$\int_0^\infty \left\{ \left| r''(u)/r^{\frac{3}{2}}(u) - \frac{2k+3}{2(k+1)}[r'(u)]^2/r^{\frac{5}{2}}(u) \right| \right\} du < \infty, \tag{3.41}$$

$$\int_0^\infty \left\{ \left| r''(u)/r^{\frac{k+3}{2(k+1)}}(u) - \frac{2k+3}{2(k+1)}[r'(u)]^2/r^{\frac{3k+5}{2(k+1)}}(u) \right| \right\} du < \infty, \tag{3.42}$$

$$|r'(t)/r^{\frac{3}{2}}(t)| \le N_1, \tag{3.43}$$

*and*

$$\int_0^t \{[r'(u)]^2/r^3(u)\}du < \infty. \tag{3.44}$$

*Then equation* (3.15) *is of the nonlinear limit–circle type if and only if*

$$\int_0^\infty [1/r^{k/(k+1)}(u)]du < \infty. \tag{3.45}$$

**Example 3.1.** Consider the equation

$$y'' + t^\sigma y^{2k-1} = 0, \quad n \ge 1. \tag{3.46}$$

As a consequence of Corollary 3.2, this equation is of the nonlinear limit–circle type (i.e., all solutions belong to $L^{2k}$), if and only if

$$\sigma > 1 + 1/k.$$

### 3.2.5. The Superlinear Forced Equation

Here, we present limit–point/limit–circle criteria for the second order forced equation with a general nonlinear term

$$(a(t)y')' + r(t)f(y) = e(t). \tag{3.47}$$

The integrability result presented here will insure that all solutions of equation (3.47) satisfy both (3.2) and (2.12).

Let $F$ be defined as in (2.13). We assume that there exist positive constants $K$ and $k$ and nonnegative constants $A$, $B$ and $C$ such that

$$K \geq 2(k+1), \tag{3.48}$$

$$0 \leq yf(y)/K - kF(y)/(k+1) \leq BF(y), \tag{3.49}$$

and

$$y^2/2 \leq AF(y) + C. \tag{3.50}$$

As in Section 3.2.1, to simplify the notation in what follows, we let $\alpha = 1/2(k+1)$ and $\beta = (2k+1)/2(k+1)$.

**Theorem 3.6.** *In addition to conditions* (3.10) *and* (3.48)–(3.50), *assume that*

$$\int_0^\infty [(a(u)r(u))'_-/a(u)r(u)]du < \infty, \tag{3.51}$$

$$\int_0^\infty \left| \{(a(u)r(u))'/a^\alpha(u)r^{\alpha+1}(u)\}' \right.$$
$$\left. -(1/k - \alpha)\left[(a(u)r(u))'\right]^2/[a^{\alpha+1}(u)r^{\alpha+2}(u)] \right| du < \infty, \tag{3.52}$$

*and*

$$\int_0^\infty \left[ |e(u)|/\left(a(u)r(u)\right)^\alpha \right] < \infty. \tag{3.53}$$

*Then equation* (3.47) *is of the nonlinear limit–circle type, i.e., any solution y of* (3.47) *satisfies* (3.2).

The proof of the above theorem can be found in Graef [59, Theorem 1]. From the proof of this theorem, which we will not give here, it can be seen that (2.12) is also satisfied for solutions of this equation. We should also note that it is possible to obtain results for equation (3.47) that are analogous to those in Sections 3.2.3–3.2.4.

A result giving necessary and sufficient conditions for the equation

$$(a(t)y')' + r(t)y^{2k-1} = e(t)$$

to be of the nonlinear limit–circle type can be found in [59, Theorem 12].

**Remark 3.5.** Theorem 3.6 can be extended to equations with more general types of perturbation terms. For example, if the function $e$ on the right-hand side of equation (3.47) is replaced by $e(t, y)$ satisfying

$$|e(t, y)| \leq h(t) F^{\frac{1}{2}}(y) + k(t),$$

then, under appropriate growth conditions on the functions $h$ and $k$, we can still obtain the conclusion of Theorem 3.6. Such results can be found in the papers of Graef and Spikes [65, Theorem 1], Spikes [104, Theorems 6 and 7], and Spikes [105, Theorem 1 and Corollary 2].

## 3.3. The Sublinear Equation

Now, we consider the sublinear equation

$$(a(t)y')' + r(t)y^\gamma = 0, \tag{3.54}$$

where $\gamma$ is the ratio of two odd positive integers with $0 < \gamma \leq 1$.

### 3.3.1. Limit–Circle Criteria

Observe that $y$ being a nonlinear limit–circle type solution of equation (3.54) means

$$\int_0^\infty y^{\gamma+1}(u) du < \infty. \tag{3.55}$$

Since $\gamma$ is the ratio of two odd positive integers, say

$$\gamma = \frac{2M - 1}{2N - 1},$$

where $M$ and $N$ are positive integers, we can write

$$\gamma = 2k - 1 \text{ where } k = \frac{M + N - 1}{2N - 1}.$$

Next, we let

$$\alpha = 1/2(k + 1) \text{ and } \beta = (2k + 1)/2(k + 1),$$

and define

$$s = \int_0^t [r^\alpha(u)/a^\beta(u)] du, \quad x(s) = y(t). \tag{3.56}$$

Making the transformation (3.56), equation (3.54) becomes

$$\ddot{x} + \alpha p(t)\dot{x} + P(t)x^\gamma = 0 \tag{3.57}$$

where

$$p(t) = (a(t)r(t))'/a^{\alpha}(t)r^{\alpha+1}(t) \text{ and } P(t) = (a(t)r(t))^{k/(k+1)}.$$

Note that although the shape of the transformation (3.56) is exactly the same as (3.5), $k$ is not an integer here. In fact, $1/2 < k \leq 1$.

We begin with a boundedness result.

**Theorem 3.7.** *If* (3.51) *holds, then every solution of* (3.54) *is bounded.*

*Proof.* Write (3.54) as the system

$$y' = w, \quad w' = (-a'(t)w - r(t)y^{\gamma})/a(t), \tag{3.58}$$

and define

$$V(t) = a(t)w^2(t)/2r(t) + y^{\gamma+1}(t)/(\gamma + 1).$$

Then,

$$V'(t) = -w^2(a(t)r(t))'/2r^2(t) \leq [(a(t)r(t))'_-/a(t)r(t)]V(t).$$

Gronwall's inequality and (3.51) imply that $V(t)$ is bounded, so $y(t)$ is bounded.
□

To prove our main limit–circle result, we write (3.57) as the system

$$\dot{x} = z - \alpha p(t)x,$$
$$\dot{z} = \alpha \dot{p}(t)x - P(t)x^{\gamma}.$$

**Theorem 3.8.** *Assume that* (3.51) *holds and*

$$\int_0^{\infty} \left| \{(a(u)r(u))'/a^{1/2}(u)r^{3/2}(u)\}' \right.$$
$$\left. + (1/2 - \alpha) [(a(u)r(u))']^2/a^{3/2}(u)r^{5/2}(u) \right| du < \infty. \tag{3.59}$$

*If*

$$\int_0^{\infty} [1/(a(u)r(u))^{k/(k+1)}]du < \infty, \tag{3.60}$$

*then equation* (3.54) *is of the nonlinear limit–circle type, i.e., any solution* $y(t)$ *of* (3.54) *satisfies* (3.55).

*Proof.* Define
$$V(x, z, s) = z^2/2 + P(t)x^{\gamma+1}/(\gamma + 1);$$

then

$$
\begin{aligned}
\dot{V} &= \alpha \dot{p}(t)xz - P(t)x^\gamma z + P(t)x^\gamma [z - \alpha p(t)x] \\
&\quad + \dot{P}(t)x^{\gamma+1}/(\gamma + 1) \\
&= \alpha \dot{p}(t)xz + [\dot{P}(t)/(\gamma + 1) - \alpha p(t)P(t)]x^{\gamma+1} \\
&= \alpha \dot{p}(t)xz.
\end{aligned}
$$

Now,

$$
\begin{aligned}
|\dot{p}(t)xz| &= \left[|\dot{p}(t)||x|^{(1-\gamma)/2}/(a(t)r(t))^{k/2(k+1)}\right] \\
&\quad \times \left[(a(t)r(t))^{k/2(k+1)}|x|^{(\gamma+1)/2}|z|\right] \\
&\le \left[|\dot{p}(t)||x|^{(1-\gamma)/2}/(a(t)r(t))^{k/2(k+1)}\right] \\
&\quad \times \left[(a(t)r(t))^{k/(k+1)}x^{\gamma+1} + z^2\right]/2.
\end{aligned}
$$

Theorem 3.7 implies that $y(t) = x(s)$ is bounded, so

$$\dot{V} \le K_1 \left[|\dot{p}(t)|/(a(t)r(t))^{k/2(k+1)}\right] V$$

for some constant $K_1 > 0$. Now $\dot{p}(t) = p'(t)a^\beta(t)/r^\alpha(t)$, and

$$
\begin{aligned}
&p'(t)/(a(t)r(t))^{k/2(k+1)} \\
&= (a(t)r(t))''/a^{1/2}(t)r^{3/2}(t) \\
&\quad - \alpha[(a(t)r(t))']^2/a^{3/2}(t)r^{5/2}(t) - (a(t)r(t))'r'(t)/a^{1/2}(t)r^{5/2}(t) \\
&= \left\{(a(t)r(t))'/a^{1/2}(t)r^{3/2}(t)\right\}' \\
&\quad + (1/2 - \alpha)[(a(t)r(t))']^2/a^{3/2}(t)r^{5/2}(t).
\end{aligned}
$$

If we let $\tau(s)$ denote the inverse function of $s(t)$, we have

$$
\begin{aligned}
&\int_0^s |\dot{p}(\tau(v))|/(a(\tau(v))r(\tau(v)))^{k/2(k+1)} \\
&= \int_0^s |p'(\tau(v))a^\beta(\tau(v))/r^\alpha(\tau(v))|/(a(\tau(v))r(\tau(v)))^{k/2(k+1)}dv \\
&= \int_0^t \left|\left\{(a(u)r(u))'/a^{1/2}(u)r^{3/2}(u)\right\}' \right.\\
&\quad \left. + (1/2 - \alpha)[(a(u)r(u))']^2/a^{3/2}(u)r^{5/2}(u)\right| du,
\end{aligned}
$$

which converges by (3.59). Hence, integrating $\dot{V}(s)$, applying Gronwall's inequality, and using condition (3.59) shows that $V(s)$ is bounded, so

$$P(t)x^{\gamma+1}(s)/(\gamma + 1) = (a(t)r(t))^{k/(k+1)}y^{\gamma+1}(t)/(\gamma + 1) \le K_2$$

for some constant $K_2 > 0$. Condition (3.60) then implies that $y(t)$ is of the nonlinear limit–circle type. $\qquad\qquad\qquad\qquad\qquad\qquad\qquad\qquad\qquad$ $\square$

**Remark 3.6.** If $\gamma = 1$ (equation (3.54) is linear), then $k = 1$ and so $\alpha = 1/4$. In this case (3.59) is the condition (1.9) of Dunford and Schwartz. Also, if $\gamma = 1$, reconstructing $V$ from $\dot{V}$ does not require that $y$ be bounded, so condition (3.51) in Theorem 3.7 is not needed. Hence, we have exactly Theorem 1.3 above of Dunford and Schwartz.

Next, we give a necessary condition for the sublinear equation to be of the nonlinear limit–circle type. It is analogous to Theorem 3.2 for superlinear equations.

**Theorem 3.9.** *Suppose* (3.29) *and* (3.51) *hold. If* $y(t)$ *is a nonlinear limit–circle type solution of* (3.54), *then*

$$\int_0^\infty \{a(u)[y'(u)]^2/r(u)\}du < \infty. \qquad (3.61)$$

*Proof.* Let $y(t)$ be a nonlinear limit–circle solution of (3.54); then $y(t)$ is bounded by Theorem 3.7. Multiply (3.54) by $y(t)/r(t)$, note that

$$(a(t)y')'y = (a(t)y'y)' - a(t)[y']^2,$$

and integrate by parts to obtain

$$a(t)y'(t)y(t)/r(t) - a(t_1)y'(t_1)y(t_1)/r(t_1)$$
$$+ \int_{t_1}^t [a(u)y'(u)y(u)r'(u)/r^2(u)]du + \int_{t_1}^t y^{\gamma+1}(u)du$$
$$- \int_{t_1}^t \{a(u)[y'(u)]^2/r(u)\}du = 0 \quad (3.62)$$

for any $t_1 \ge 0$. Schwarz's inequality, the fact that $y(t)$ is bounded, and (3.29) imply

$$\left| \int_{t_1}^t [a(u)y'(u)y(u)r'(u)/r^2(u)] du \right|$$

$$\leq \left[ \int_{t_1}^t \{a(u)[y'(u)]^2/r(u)\} du \right]^{\frac{1}{2}} \left[ \int_{t_1}^t \{a(u)y^2(u)[r'(u)]^2/r^3(u)\} du \right]^{\frac{1}{2}}$$

$$\leq K_1 \left[ \int_{t_1}^t \{a(u)[y'(u)]^2/r(u)\} du \right]^{\frac{1}{2}}.$$

If $y(t)$ is not eventually monotonic, let $\{t_j\} \to \infty$ be an increasing sequence of zeros of $y'(t)$. Then, (3.62) implies

$$K_1 H^{\frac{1}{2}}(t_j) + K_2 \geq H(t_j)$$

where

$$H(t) = \int_{t_1}^t \{a(u)[y'(u)]^2/r(u)\} du$$

and $K_2 > 0$ is a constant. This implies $H(t_j) \leq K_3 < \infty$ for all $j$ and some $K_3 > 0$, so (3.61) holds.

If $y(t)$ is eventually monotonic, then $y(t)y'(t) \leq 0$ for $t \geq t_1$ for large $t_1 \geq 0$. Using this in (3.62) and repeating the type of argument used above, we again obtain that (3.61) holds. $\qquad\square$

### 3.3.2. Limit–Point Criteria

**Theorem 3.10.** *Suppose conditions* (3.27), (3.29), (3.51), *and* (3.59) *hold. If*

$$\int_0^\infty [1/(a(u)r(u))^{k/(k+1)}] du = \infty, \tag{3.63}$$

*then every nontrivial solution of* (3.54) *is of the nonlinear limit–point type.*

*Proof.* Proceeding as in the proof of Theorem 3.8, we have $\dot{V}(s) = \alpha \dot{p}(t)xz$. Let $y(t) = x(s)$ be any nontrivial solution of (3.54) with $y(t_1) = x(s(t_1)) = x(s_1) \neq 0$. Theorem 3.7 implies that $x(s)$ is bounded, so $|x(s)|^{(1-\gamma)/2} \leq K_1$ for some $K_1 > 0$. Hence,

$$\dot{V}(s) \geq -\alpha|\dot{p}(t)||x(s)|^{(1-\gamma)/2}|x(s)|^{(1+\gamma)/2}|z(s)|$$
$$\geq -\alpha K_1 \left[ |\dot{p}(t)|/(a(t)r(t))^{k/2(k+1)} \right]$$
$$\times \left[ (a(t)r(t))^{k/2(k+1)}x^{\gamma+1}(s) + z^2(s) \right]/2$$
$$\geq -K_2 \left[ |\dot{p}(t)|/(a(t)r(t))^{k/2(k+1)} \right] V.$$

If we let

$$H(t) = K_2|\dot{p}(t)|/(a(t)r(t))^{k/2(k+1)},$$

then we can write

$$\dot{V} + H(t)V \geq 0,$$

so

$$\frac{d}{ds}\left(V(s)\exp\int_{s_1}^{s} H(\tau(\xi))d\xi\right) \geq 0.$$

Integrating, we obtain

$$V(s)\exp\int_{s_1}^{s} H(\tau(\xi))d\xi \geq V(s_1).$$

Condition (3.59) implies

$$\int_{s_1}^{\infty} H(\tau(\xi))d\xi < \infty,$$

and since $V(s_1) > 0$, we have

$$V(s) \geq K_3 > 0 \quad \text{for } s \geq s_1.$$

Dividing by $(a(t)r(t))^{k/(k+1)}$ and rewriting the result in terms of $t$, we obtain

$$a(t)[y'(t)]^2/2r(t) + \alpha(a(t)r(t))'y(t)y'(t)/r^2(t)$$
$$+ \alpha^2[(a(t)r(t))']^2 y^2(t)/2a(t)r^3(t) + y^{\gamma+1}(t)/(\gamma+1)$$
$$\geq K_3/(a(t)r(t))^{k/(k+1)}. \quad (3.64)$$

If $y(t)$ is a limit–circle type solution, then

$$\int_{t_1}^{\infty} y^{\gamma+1}(u)du < \infty \text{ and } \int_{t_1}^{\infty} \{a(u)[y'(u)]^2/r(u)\}du < \infty.$$

Also, by Theorem 3.7, $y(t)$ is bounded, so (3.27) implies

$$\int_{t_1}^{\infty} \left\{[(a(u)r(u))']^2 y^2(u)/2a(u)r^3(u)\right\} du < \infty.$$

Schwarz's inequality yields

$$\left|\int_{t_1}^{t} \{(a(u)r(u))'y(u)y'(u)/r^2(u)\}du\right|$$
$$\leq \left[\int_{t_1}^{t} \{[(a(u)r(u))']^2 y^2(u)/a(u)r^3(u)\}du\right]^{1/2}$$
$$\times \left[\int_{t_1}^{t} \{a(u)[y'(u)]^2/r(u)\}du\right]^{1/2},$$

so the integral of the second term on the left-hand side of (3.64) converges. Thus, integrating (3.64) and applying (3.63) yields a contradiction. Therefore, $y(t)$ is a nonlinear limit–point type solution. ☐

**Theorem 3.11.** *Suppose that (3.29) and (3.51) hold. If*

$$\int_0^\infty [1/a(u)r(u)]du = \infty,$$

*then (3.54) is of the nonlinear limit–point type.*

*Proof.* Again we write (3.54) as the system (3.58) and define

$$V(y, w, t) = V(t) = a^2(t)w^2(t)/2 + a(t)r(t)y^{\gamma+1}(t)/(\gamma + 1).$$

Then,

$$V'(t) = (a(t)r(t))'y^{\gamma+1}/(\gamma + 1) \geq -[(a(t)r(t))'_-/a(t)r(t)]V(t),$$

so

$$\left(V(t)\exp\int_0^t [(a(u)r(u))'_-/a(u)r(u)]du\right)' \geq 0.$$

Now, (3.51) implies

$$\exp\int_0^t [(a(u)r(u))'_-/a(u)r(u)]du \leq K_1$$

for some $K_1 > 0$, so

$$V(t) \geq V(0)/K_1.$$

If $y(t)$ is any solution of (3.54) such that $V(y(0), w(0), 0) > 2K_1$, then $V(t) \geq 2$ and we have

$$\int_0^t [V(u)/a(u)r(u)]du \geq \int_0^t [2/a(u)r(u)]du \to \infty$$

as $t \to \infty$. Thus,

$$\int_0^t \left\{a(u)[y'(u)]^2/2r(u) + y^{\gamma+1}(u)/(\gamma + 1)\right\} du \to \infty$$

as $t \to \infty$, which is a contradiction. ☐

### 3.3.3. Necessary and Sufficient Conditions

Combining Theorems 3.8 and 3.10 we have the following result.

**Theorem 3.12.** *Assume that* (3.27), (3.29), (3.51), *and* (3.59) *hold. Then* (3.54) *is of the nonlinear limit–circle type if and only if*

$$\int_0^\infty [1/(a(u)r(u))^{k/(k+1)}]du < \infty.$$

If we apply this result to the case $a(t) \equiv 1$, we have the following corollary.

**Corollary 3.3.** *Assume that*

$$\int_0^\infty [r'(u)_-/r(u)]du < \infty,$$

$$\int_0^\infty \left| \{r'(u)/r^{3/2}(u)\}' + \frac{k}{2(k+1)} [r'(u)]^2 /r^{5/2}(u) \right| du < \infty,$$

*and*

$$\int_0^t \{[r'(u)]^2/r^3(u)\}du < \infty.$$

*Then the equation*

$$y'' + r(t)y^\gamma = 0, \quad 0 < \gamma \le 1, \tag{3.65}$$

*is of the nonlinear limit–circle type if and only if*

$$\int_0^\infty [1/r^{k/(k+1)}(u)]du < \infty.$$

When specialized to the equation

$$y'' + t^\sigma y^\gamma = 0, \tag{3.66}$$

we have that (3.66) is of the nonlinear limit–circle type if and only if

$$\sigma > 1 + 1/k = 1 + 2/(\gamma + 1).$$

This agrees with what is known from asymptotic integrations of this equation (see, for example, Bellman [21, p. 163]), and shows that our results are sharp.

### 3.3.4. The Sublinear Forced Equation

Finally, we present nonlinear limit–circle criteria for the sublinear forced equation

$$(a(t)y')' + r(t)y^\gamma = e(t), \tag{3.67}$$

where $\gamma$ is the ratio of two odd positive integers with $0 < \gamma \leq 1$ and $e(t)$ is continuous.

**Theorem 3.13.** [60, Theorem 2] *Suppose that conditions* (3.51), (3.59), *and* (3.60) *hold and*

$$\int_0^\infty \left[ |e(u)|/(a(u)r(u))^\alpha \right] du < \infty. \tag{3.68}$$

*Then equation* (3.67) *is of the nonlinear limit–circle type, i.e., any solution $y(t)$ of* (3.67) *satisfies* (3.55).

## 3.4. Equations with $r(t) \leq 0$

In this section, we continue our discussion of second order nonlinear equations but with $r$ having the opposite sign from that of equation (3.1) considered in Sections 3.2 and 3.3 above.

We consider the second order nonlinear differential equation

$$y'' = r(t)f(y) \tag{3.69}$$

where $r : \mathbb{R}_+ \to \mathbb{R}$ and $f : \mathbb{R} \to \mathbb{R}$ are continuous, $r(t) \geq 0$, and $yf(y) \geq 0$ for all $y$. It is known (see [14]) that if $f(y) \equiv y$, then there exists a solution $y$ of (3.69) that does not belong to $L^2$. But if $f$ is strongly nonlinear, the situation is considerably different. We only consider those solutions of (3.4) that are defined on $[0, \tau)$, $\tau \leq \infty$, and that can not be defined at $\tau$ in case $\tau < \infty$. As before, if $\tau = \infty$, the solution is called *continuable*, and if $\tau < \infty$, it is said to be *noncontinuable*. In view of what we discussed earlier, the following definition should come as no surprise.

**Definition 3.1.** A continuable solution $y$ of (3.69) is said to be of the *nonlinear limit–circle type* if

$$\int_0^\infty y(t)f(y(t))dt < \infty;$$

otherwise, it is said to be of the *nonlinear limit–point type*. We will say that equation (3.69) is of the *nonlinear limit–circle type* if every continuable solution is of the nonlinear limit–circle type. If there exists at least one continuable solution $y$ of (3.69) that is of the nonlinear limit–point type, then equation (3.69) is said to be of the *nonlinear limit–point type*.

The following theorem is a special case of Theorem 8.1 below. We include it here both for completeness as well as to motivate the results in the following sections.

**Theorem 3.14.** *If there exist constants $M > 0$ and $M_1 > 0$ such that*

$$\frac{1}{|x|} \leq |f(x)| \leq M_1|x| \tag{3.70}$$

*for $x \geq M$ or $x \leq -M$, then (3.69) is of the nonlinear limit–point type.*

In the remainder of this chapter, we examine equation (3.69) in the case where (3.70) does not hold.

### 3.4.1. Nonlinear Limit–Point Results

We begin with a lemma that categorizes the behavior of solutions of equation (3.69).

**Lemma 3.2.** *If $y$ is a solution of (3.69) defined on $[0, \tau)$, $\tau \leq \infty$, then either*

$$y(t)y'(t) \leq 0 \quad on \quad R_+, \tag{3.71}$$

*or*

$$y(t)y'(t) > 0 \quad on \quad [t_0, \tau), \quad t_0 > 0, \quad and \quad \lim_{t \to \tau^-} |y(t)| = \infty. \tag{3.72}$$

It is not difficult to see that solutions of the types described in (3.71) and (3.72) do exist. For example, a solution $y$ such that $y(0) = y'(0) = 1$ is of the type (3.72). The existence of solutions satisfying (3.71) follows from [80, Theorem 13.1].

Our first result here gives an easily verifiable integral condition on $r$ as a criteria for equation (3.69) to be of the limit–point type. The proof is based on Corollary 8.2 in [80] which guarantees the existence of a continuable solution $y$ of (3.69) satisfying $\lim_{t \to \infty} y(t) = 1$.

**Theorem 3.15.** *If*

$$\int_0^\infty tr(t)dt < \infty, \tag{3.73}$$

*then equation (3.14) is of the nonlinear limit–point type.*

The following result gives sufficient conditions for solutions of (3.69) that satisfy (3.71) to be of the limit–circle type.

**Proposition 3.1.** *A solution y of (3.69) satisfying (3.71) is of the nonlinear limit–circle type if either of the following conditions holds:*

(i)     *There exists $c > 0$ such that*

$$r(t) \geq \frac{c}{t} \quad \text{for all large } t; \tag{3.74}$$

(ii)    *There exist $\varepsilon > 0$ and $\sigma > 0$ such that*

$$|f(x)| \geq |x| \quad \text{for} \quad |x| \leq \varepsilon \tag{3.75}$$

*and*

$$r(t) \geq \frac{1+\sigma}{t^2} \quad \text{for all large } t. \tag{3.76}$$

*Proof.* To prove the first part, define the function $F$ by

$$F(t) = ty(t)y'(t) - \frac{y^2(t)}{2}, \quad t \in R_+.$$

Along solutions of equation (3.69), we have

$$F'(t) = tr(t)y(t)f(y(t)) + ty'^2(t) \geq tr(t)y(t)f(y(t)) \geq 0, \ t \in R_+. \tag{3.77}$$

Thus, $F$ is nondecreasing, and since $F(t) \leq 0$ on $R_+$, we have that $F$ is bounded. Hence,

$$\infty > \int_0^\infty F'(t)dt \geq \int_0^\infty tr(t)y(t)f(y(t))dt \geq c \int_T^\infty y(t)f(y(t))dt,$$

so $y$ is of the nonlinear limit–circle type.

Now consider case (ii) and let (3.76) hold for $t \geq T \geq 0$, and define $G$ by

$$G(t) = t^2 y(t)y'(t) - ty^2(t)$$

for $t \in R_+$. Then, $G(t) \leq 0$ on $R_+$ and

$$G'(t) = t^2(y'(t))^2 + t^2 r(t)y(t)f(y(t)) - y^2(t), \ t \in R_+.$$

By [80, Theorem 13.2], (3.76) implies $\lim\limits_{t \to \infty} y(t) = 0$, so $|y(t)| \leq \varepsilon$ for $t \geq T_1$ for some $T_1 \geq T$. Now (3.75) and (3.76) imply

$$G'(t) \geq (t^2 r(t) - 1)y(t)f(y(t)) \geq \sigma y(t)f(y(t)) \geq 0$$

for $t \geq T_1$. Thus, $G$ is nondecreasing and nonpositive for $t \geq T_1$, and so it is bounded. Hence,

$$\infty > \int_{T_1}^\infty G'(t)dt \geq \sigma \int_{T_1}^\infty y(t)f(y(t))dt,$$

and the conclusion of the proposition follows.      □

Conditions (3.75) and (3.76) are really quite sharp as the following example will show.

**Example 3.2.** The equation

$$y'' = t^{-2}y$$

has the solution $y(t) = t^{(1+\sqrt{5})/2}$ defined on $[1, \infty)$ with

$$\int_1^\infty y(t)f(y(t))dt = \int_1^\infty y^2(t)dt = \infty,$$

i.e., $y$ is a nonlinear limit–point type solution. Note that condition (3.75) is satisfied for this equation. This shows that the constant $1 + \sigma$ in (3.76) is quite sharp. On the other hand, the equation

$$y'' = \frac{1}{2+\gamma}\left(\frac{1}{2+\gamma} + 1\right)t^{-\frac{4+\gamma}{2+\gamma}}|y|^{1+\gamma}\,\text{sgn}\,y, \quad \gamma > 0, \quad t \geq 1,$$

has the solution $y = t^{-1/(2+\gamma)}$. Here, (3.76) is satisfied, but

$$\int_1^\infty y(t)f(y(t))dt = \int_1^\infty [y(t)]^{2+\gamma}dt = \int_1^\infty t^{-1}dt = \infty,$$

so $y$ is of the nonlinear limit–point type. Thus, (3.75) cannot be replaced by the condition

$$|f(x)| \geq |x|^{1+\gamma}, \quad \gamma > 0, \quad \text{for} \quad |x| \leq \varepsilon.$$

### 3.4.2. Nonlinear Limit–Circle Results

Based on Proposition 3.1 we have the following limit–circle result.

**Theorem 3.16.** *In addition to the hypotheses of Proposition 3.1, assume that there exist $\lambda > 1$ and $M > 0$ such that*

$$|f(x)| \geq |x|^\lambda \quad \text{for} \quad |x| \geq M. \tag{3.78}$$

*Then equation (3.69) is of the nonlinear limit–circle type.*

**Proof.** In view of Lemma 3.2 and Proposition 3.1, we only need to consider the case where (3.72) holds. Let $y$ be a continuable solution of (3.69) satisfying (3.72). If either case (i) or (ii) of Proposition 3.1 holds, then there exist constants $C_1 > 0$ and $M_1 \geq 0$ such that

$$r(t) \geq \frac{C_1}{t^2} \quad \text{for} \quad t \geq M_1.$$

Theorem 11.4 in [80] then yields a contradiction to the continuability of $y$. $\qquad\square$

We wish to point out that in Theorem 3.16, the right-hand portion of inequality
(3.70) in Theorem 3.14 does not hold. For our final result in this section, we
examine the situation where the left-hand inequality in (3.70) fails to hold, i.e.,
$|f(x)|$ is small for large $|x|$.

**Theorem 3.17.** *In addition to the hypotheses of Proposition 3.1, assume that one
of the following conditions hold:*

(i) *There exist $M_2 > 0$ and $\varepsilon > 0$ such that*

$$|f(x)| \le \frac{1}{|x|^{2+\varepsilon}} \quad for \quad |x| \ge M_2;$$

(ii) *There exist constants $1 < \mu \le \nu < \infty$ and $M_3 \ge 1$ such that*

$$\frac{1}{|x|^\nu} \le |f(x)| \le \frac{1}{|x|^\mu} \quad for \quad |x| \ge M_3$$

*and*

$$\int_0^\infty \left( \int_0^t \int_0^s r(\tau)d\tau ds \right)^{\frac{1-\mu}{1+\nu}} dt < \infty.$$

*Then all solutions of equation (3.69) are continuable and equation (3.69) is of the
nonlinear limit–circle type.*

As a consequence of Theorem 3.17, the following example shows that the left
hand inequality in (3.70) is sharp.

**Example 3.3.** Consider the differential equation

$$y'' = t^{(2+\rho)/\rho} f(y), \quad t \ge 1,$$

where $f$ is given by

$$f(x) = \begin{cases} x & \text{if } |x| \le 1, \\ \dfrac{\operatorname{sgn} x}{|x|^{1+\rho}} & \text{if } |x| > 1. \end{cases}$$

Then, Theorem 3.17(ii) holds with $\mu = \nu = 1 + \rho$, and so this equation is of the
nonlinear limit–circle type. This shows that the first inequality in (3.70) can not
be replaced by the weaker condition $1/|x|^\omega \le |f(x)|$ with $\omega > 1$.

The following necessary condition for a solution of equation (3.69) to be of
the nonlinear limit–circle type is essentially identical to Theorem 3.2 above for
equation (3.4).

**Theorem 3.18.** [66, Theorem 1] *Suppose that conditions* (3.28)–(3.29) *hold, and there are constants* $M_4 > 0$ *and* $M_5 > 0$ *such that*

$$u^2 \leq M_4 u f(u) + M_5. \tag{3.79}$$

*If* $y$ *is a nonlinear limit–circle type solution of equation* (3.69), *then*

$$\int_0^\infty \{a(u)[y'(u)]^2/r(u)\}du < \infty. \tag{3.80}$$

Using the above result, we can then give a sufficient condition for equation (3.69) to be of the nonlinear limit–point type.

**Theorem 3.19.** [66, Theorem 5] *If conditions* (3.28), (3.29), *and* (3.79) *hold and* $(a(t)r(t))' \leq 0$, *then equation* (3.69) *is of the nonlinear limit–point type.*

*Proof.* Define $V(t) = F(y(t)) - a(t)[y'(t)]^2/2r(t)$ where $F(y) = \int_0^y f(u)du$ as before. Then,

$$V'(t) = \frac{a'(t)[y'(t)]^2}{r(t)} - \frac{[y'(t)]^2[r(t)a'(t) - a(t)r'(t)]}{2r^2(t)} = \frac{(a(t)r(t))'[y'(t)]^2}{2r^2(t)} \leq 0.$$

Choose a solution $y(t)$ of equation (3.69) with $y(0) = y_0$ and $y'(0) = y_1$ satisfying $V(0) = -c < 0$. Assume that $y$ is a nonlinear limit–circle type solution. Integrating $V'(t)$, we obtain $V(t) \leq V(0) \leq -c$, and a second integration yields

$$-\int_0^t [a(u)[y'(u)]^2/2r(u)]du \leq \int_0^t V(u)du \leq -ct \to -\infty$$

as $t \to \infty$. This contradicts (3.80) and completes the proof of the theorem. $\square$

A result on the relationship between the nonlinear limit–circle property and the convergence to zero of solutions of equation (3.69) is the following.

**Theorem 3.20.** [66, Corollary 4] *Let* $f(u)$ *be bounded away from zero if* $u$ *is bounded away from zero. If* $y$ *is a nonlinear limit–circle type solution of equation* (3.69), *then* $y(t) \to 0$ *as* $t \to \infty$ *and* $y(t)y'(t) < 0$ *for all large* $t$.

**Open Problems.**

**Problem 3.1.** *Prove Theorem 1.11 above for nonlinear equations.*

**Problem 3.2.** *Prove either part of Theorem 1.9 for nonlinear equations.*

**Problem 3.3.** *Kroopnick [86, 87] obtained some integrability results for Liénard type equations. Extend the limit–point/limit–circle results in this chapter to second order equations with damping terms.*

**Notes.** The results in Sections 3.1.1 to 3.1.4 are specializations to the equation (3.4) of results in Graef [59]. The results in Sections 3.2.1 to 3.2.3 can be found in [60]. Except where noted, the results in Section 3.4 are based on the paper [19] of Bartušek and Graef.

# Chapter 4

# Some Early Limit–Point and Limit–Circle Results

In the introduction to Chapter 3, we made reference to some papers that give sufficient conditions for the existence of limit–point solutions to nonlinear equations. In this chapter, we will describe some of these results and relate them to the work of Dunford and Schwartz as well as to our results in Chapter 3. In the first section, we discuss an important, and one of the earliest, results of this type. It was obtained by Wintner [115] for linear equations.

## 4.1. Wintner's Result

Wintner [115] considered the second order linear equation

$$y'' + r(t)y = 0 \tag{4.1}$$

where $r : \mathbb{R}_+ \to \mathbb{R}$ is continuous and he gave a sufficient condition for the nonexistence of $L^2$ solutions. His result is as follows.

**Theorem 4.1.** *If*

$$\int_0^\infty u^3 r^2(u)du < \infty, \tag{4.2}$$

*then equation (4.1) has no nontrivial solutions belonging to $L^2$, i.e., equation (4.1) is of the limit–point type.*

As Wintner points out, this result is the best possible in some sense as the next example shows.

**Example 4.1.** The equation

$$y'' - \frac{\alpha(\alpha + 1)}{t^2} y = 0 \tag{4.3}$$

satisfies

$$\int_0^\infty u^{3-\epsilon} |r(u)|^2 du < \infty \text{ and } \int_0^\infty u^3 |r(u)|^{2+\epsilon} du < \infty \tag{4.4}$$

and has the solution $y(t) = t^{-\alpha}$. Note that condition (4.2) does not hold. Clearly, this solution belongs to $L^2[0, \infty)$ if and only if $\alpha > 1/2$. Observe that the results of Dunford and Schwartz can only apply if $-1 < \alpha < 0$ since they need $r(t) > 0$. Yet while condition (1.11) of Theorem 1.4 holds, condition (1.9) (see (1.15)) does not. Hence, we can not conclude anything from their result.

In proving Theorem 4.1, Wintner first proves the following lemma that is somewhat interesting in its own right.

**Lemma 4.1.** *If*

$$\int_0^\infty u^2 r^2(u) du < \infty \tag{4.5}$$

*and equation (4.1) has a nontrivial solution $y \in L^2[0, \infty)$, then*

$$\int_0^\infty |y'(u)| du < \infty. \tag{4.6}$$

It follows from Lemma 4.1 that $y(t) \to 0$ as $t \to \infty$ since (4.6) implies that the total variation of $y$ is finite and thus $y$ must have a limit.

## 4.2. Early Results on Higher Order Linear Equations

No discussion of the limit–point/limit–circle problem would be complete without some mention of the early work of Naimark and Fedoryuk on higher order linear equations. We refer the reader to the English translation [94] of Naimark's 1954 monograph; some of Fedoryuk's papers are available in translation (see, for example, [52, 53, 54]).

### 4.2.1. Naimark's Results

We consider the differential expression

$$\ell(y) \equiv \sum_{i=0}^n (-1)^i \left( p_i(t) y^{(i)} \right)^{(i)} = (-1)^n (p_n(t) y^{(n)})^{(n)}$$

$$+ (-1)^{n-1} (p_{n-1}(t) y^{(n-1)})^{(n-1)} + \cdots + (-1)(p_1(t) y^{(1)})^{(1)} + p_0(t) y \tag{4.7}$$

for $t \in \mathbb{R}_+$, where $p_n^{-1}, p_{n-1}, \ldots, p_0 \in L_{\text{loc}}(\mathbb{R}_+)$, and $p_0(t) > 0$. The minimal operator associated with this expression will be denoted by $L_0$. The types of conditions used by Naimark include:

$$\frac{p_n'}{p_n}, \frac{p_{n-1}}{p_0^{1/2n}}, \frac{p_{n-2}}{p_0^{3/2n}}, \frac{p_{n-3}}{p_0^{5/2n}}, \ldots, \frac{p_1}{p_0^{(2n-3)/2n}} \in L(\mathbb{R}_+); \tag{4.8}$$

$$p_0' = \mathcal{O}(|p_0|^{\alpha}), \quad 0 < \alpha < 1 + \frac{1}{2n} \quad \text{as} \quad t \to \infty; \tag{4.9}$$

$$\frac{p_0''}{p_0^{1+1/2n}}, \frac{(p_0')^2}{p_0^{2+1/2n}} \in L(\mathbb{R}_+). \tag{4.10}$$

As a special case of (4.7), we have

$$\ell(y) \equiv (-1)^k y^{(k)} + r(t)y. \tag{4.11}$$

One of Naimark's most interesting results, and one which has hypotheses similar to those used in this monograph, is the following. It is based on Theorem 4 of §23.4 in [94]. (Also see Theorem 7 in §24.3 and Theorem 8 in §24.4 of [94].)

**Theorem 4.2.** *Assume that* $|p_0(t)| \to \infty$ *as* $t \to \infty$, $p_0'$ *and* $p_0''$ *do not change signs for sufficiently large* $t$, *and conditions* (4.8) *and* (4.9) *hold.*

(i) *If* $p_0(t) \to +\infty$ *as* $t \to \infty$, *the deficiency index of the operator* $L_0$ *is* $n$.

(ii) *If* $p_0(t) \to -\infty$ *as* $t \to \infty$, *the deficiency index of the operator* $L_0$ *is* $n + 1$ *if the integral*

$$\int_0^\infty |p_0(u)|^{-1+1/2n} du \tag{4.12}$$

*converges, and the deficiency index of the operator* $L_0$ *is* $n$ *if the integral in* (4.12) *diverges.*

In proving his results, Naimark shows that condition (4.9), the fact that $p_0'$ does not change sign, and the assumption that $|p_0(t)| \to \infty$ as $t \to \infty$, imply

$$\int_0^\infty \frac{[p_0'(u)]^2}{p_0^{2+1/2n}(u)} du < \infty \quad \text{and} \quad \int_0^\infty \frac{p_0''(u)}{p_0^{1+1/2n}(u)} du < \infty. \tag{4.13}$$

Notice that if $n = 1$ and $p_0 = r$, these become

$$\int_0^\infty \frac{[r'(u)]^2}{r^{5/2}(u)} du < \infty \quad \text{and} \quad \int_0^\infty \frac{r''(u)}{r^{3/2}(u)} du < \infty,$$

which are the individual terms in the integral condition (1.15) of Dunford and Schwartz.

When we apply Theorem 4.2 to the equation (4.11) with $k = 2$, we obtain the following result.

**Corollary 4.1.** *Let $r'$ and $r''$ have constant signs and let $r'(t) = \mathcal{O}(|r(t)|^c)$ as $t \to \infty$ for $0 < c < 3/2$. If $r(t) \to +\infty$ as $t \to \infty$, then the operator $L_0$ generated by the differential expression*

$$\ell(y) \equiv -y'' + r(t)y \tag{4.14}$$

*has deficiency index 1. If $r(t) \to -\infty$ as $t \to \infty$, then the deficiency index of the operator $L_0$ is 2 if the integral*

$$\int_0^\infty |r(u)|^{-1/2} du \tag{4.15}$$

*converges, and is 1 if the integral (4.15) diverges.*

**Remark 4.1.** The second part of Corollary 4.1 was known to Titchmarsh; for example, see [110, Section 3] or [109], or see Section 1.3 above.

It is interesting to note that in proving his results for (4.7), Naimark makes use of the quantity

$$s = \int_{t_0}^t \left[ (-1)^{n-1} \frac{p_0(u)}{p_n(u)} \right]^{\frac{1}{2n}} du$$

which, for $n = 1$, reduces to the transformation (1.8) discussed in Section 1.4.

For further discussions of these results, we refer the reader to §22 – §24 in Naimark [94].

### 4.2.2. Fedoryuk's Results

In addition to considering self-adjoint differential expressions of the form (4.7), Fedoryuk often studied differential expressions in the nonself–adjoint form

$$\ell(y) \equiv \sum_{i=0}^n p_i(t) y^{(i)} = p_n(t) y^{(n)} + p_{n-1}(t) y^{(n-1)} + \ldots$$

$$+ p_1(t) y^{(1)} + p_0(t) y \tag{4.16}$$

for $t \in \mathbb{R}_+$, where $p_n(t) \equiv 1$ and $p_{n-1}, \ldots, p_0 \in L_{\text{loc}}(\mathbb{R}_+)$. The types of conditions he imposed were somewhat different than those used by Naimark. For example, Fedoryuk (see [52], [54], [53]) asked that

$$\frac{(p_0')^2}{p_0^{2+1/n}}, \frac{(p_1')^2}{p_0^{2-1/n}}, \frac{(p_2')^2}{p_0^{2-3/n}}, \cdots, \frac{(p_{n-1}')^2}{p_0^{2-(2n-3)/n}} \in L(\mathbb{R}_+) \qquad (4.17)$$

and

$$\frac{p_0''}{p_0^{1+1/n}}, \frac{p_1''}{p_0^{1}}, \frac{p_2''}{p_0^{1-1/n}}, \frac{p_3''}{p_0^{1-2/n}}, \cdots, \frac{p_{n-1}''}{p_0^{1-(n-2)/n}} \in L(\mathbb{R}_+). \qquad (4.18)$$

Due to the fact that Fedoryuk develops asymptotic formulas for the behavior of solutions, it is often necessary that he requires certain quotients of the coefficients in equations (4.7) and (4.16) to have a prescribed limiting behavior. For further details, we refer the reader to the lengthy work [53] or his monograph [54].

## 4.3. Nonlinear Limit–Point Results for Second Order Equations

In this section, we discuss some extensions of Wintner's theorem to second order nonlinear equations. We begin with a result of Suyemoto and Waltman [107]. Consider the equation

$$y'' + r(t)y^\rho = 0 \qquad (4.19)$$

where $r : \mathbb{R}_+ \to \mathbb{R}$ is continuous and $\rho \geq 1$.

**Theorem 4.3.** *If (4.2) holds, then equation (4.19) has no nontrivial solutions in the class $L^{2\rho}[0, \infty)$, i.e., every nontrivial solution $y$ of (4.19) satisfies*

$$\int_0^\infty |y(u)|^{2\rho} du = \infty. \qquad (4.20)$$

In order to relate Theorem 4.3 to our results in Chapter 3, let us consider the special case of equation (4.19) with $\rho$ an odd integer, say $\rho = 2k - 1$ where $k = 1, 2, \ldots$, and $r(t) = t^\sigma$, i.e., equation (3.24). Condition (4.2) becomes

$$\sigma < -2.$$

Clearly, this is not nearly as good as the condition

$$\sigma \leq 1 + \frac{1}{k}$$

(see Remark 3.3 above). One reason for this is that condition (4.2) does not take into account the size of the nonlinearity, i.e., the value of $k$.

Burlak [23] studied a similar problem for the equation

$$y'' + r(t)f(y) = 0 \tag{4.21}$$

where $r : \mathbb{R}_+ \to \mathbb{R}$ and $f : \mathbb{R} \to \mathbb{R}$ are continuous, but $f$ was not required to satisfy the usual sign condition $uf(u) \geq 0$ for all $u \neq 0$. Condition (4.2) was imposed on $r$ as above.

In [32], Detki also considered equation (4.21) under the assumptions that $r : \mathbb{R}_+ \to \mathbb{R}$ is continuous, $f : \mathbb{R} \to \mathbb{R}$ belongs to $C^1(-\infty, \infty)$,

$$f(0) = 0, \quad f'(0) > 0, \quad \text{and} \quad f'(u) \geq 0 \quad \text{for all} \quad u. \tag{4.22}$$

His result is the following.

**Theorem 4.4.** ([32, Theorem 1]) *If (4.2)* holds, then every nontrivial solution y *of (4.21)* satisfies

$$\int_0^\infty f^2(y(u))du = \infty. \tag{4.23}$$

Notice that in Theorem 4.4, if $f$ is linear, then the conclusion of the theorem is that no nontrivial solution $y$ satisfies

$$\int_0^\infty y^2(u)du < \infty,$$

that is, the equation is of the limit–point type. If $f(u) = u^{2k-1}$ with $k = 1, 2, \ldots$, as in equation (3.15) above, then (4.23) becomes

$$\int_0^\infty y^{4k-2}(u)du = \infty.$$

This is compared to

$$\int_0^\infty y^{2k}(u)du = \infty$$

in our definition of a nonlinear limit–point solution. Once again, we see that no advantage is made of the size of the nonlinearity in the conditions on $r(t)$.

Wong [116] studied the equation

$$y'' + g(t, y) = 0, \tag{4.24}$$

where $g : \mathbb{R}_+ \times \mathbb{R} \to \mathbb{R}$ is continuous, satisfies

$$|g(t, y)| \leq r(t)f(y),$$

and $r : \mathbb{R}_+ \to \mathbb{R}_+$ and $f : \mathbb{R} \to \mathbb{R}_+$ are continuous. One of his results that is comparable to those in this section is as follows.

**Theorem 4.5.** ([116, Corollaries 1 and 2]) *Assume that* $f(u) \neq 0$ *if* $u \neq 0$ *and there are constants* $k_1 > 0$, $k_2 > 0$, *and* $\rho \geq 1$ *such that* $\liminf_{u \to \infty} f(u) \geq k_1$, *and for some* $\epsilon > 0$ *we have*

$$|f(u)| \leq k_2 |u|^\rho \quad \text{if} \quad |u| < \epsilon \quad \text{and} \quad |u| > \frac{1}{\epsilon}.$$

*If* (4.2) *holds, then no solution of* (4.24) *belongs to* $L^{2\rho}[0, \infty)$.

## 4.4. Nonlinear Limit–Point Results for Higher Order Equations

In this section, we describe extensions of Wintner's theorem to higher order nonlinear equations. We begin with a result of Hallam [70] who considered the $n$-th order equation

$$y^{(n)} + f(t, y) = 0 \tag{4.25}$$

where $n \geq 1$, $f : \mathbb{R}_+ \times \mathbb{R} \to \mathbb{R}$ is continuous and satisfies

$$|f(t, y)| \leq r(t)|y|^\rho, \tag{4.26}$$

$r : \mathbb{R}_+ \to \mathbb{R}_+$ is continuous, and $\rho \geq 1$.

**Theorem 4.6.** ([70, Theorem 1]) *If*

$$\int_0^\infty u^{2n-1} r^2(u) du < \infty, \tag{4.27}$$

*then no nontrivial solution of* (4.25) *belongs to* $L^{2\rho}[0, \infty)$.

If $n = 2$, then condition (4.27) is exactly condition (4.2) of Wintner [115], Suyemoto and Waltman [107], Burlak [23], and Detki [32] described above. Also note that Hallam's condition (4.27) does involve the order $n$ of the equation but it *does not* take advantage of the size of the nonlinear function $f$. This is the same weakness that we noted in Theorems 4.3–4.5 above. To see that condition (4.27) cannot be replaced by the pair of conditions

$$\int_0^\infty u^{2n-1-\epsilon} |r(u)|^2 du < \infty \text{ and } \int_0^\infty u^{2n-1} |r(u)|^{2+\epsilon} du < \infty \tag{4.28}$$

with $\epsilon > 0$, we present the following example.

**Example 4.2.** Consider the equation

$$y^{(n)} - \frac{\alpha^{(n)}}{t^{n+(\rho-1)\alpha}} y = 0 \tag{4.29}$$

where $\alpha^{(n)} = \alpha(\alpha - 1)(\alpha - 2) \cdots (\alpha - n + 1)$ is the generalized factorial. This equation has the solution $y(t) = t^\alpha$. Clearly, $\alpha$ can be chosen so that (4.28) holds and $y$ belongs to $L^2[0, \infty)$.

A generalization of Wintner's and Detki results was obtained by Elias [43] who considered the equation

$$y^{(n)} + r(t)f(y) = 0 \tag{4.30}$$

with $n \geq 1$, $r : \mathbb{R}_+ \to \mathbb{R}$ and $f : \mathbb{R} \to \mathbb{R}$ continuous, $f \in C^1(\mathbb{R})$, and (4.22) holding. The main result in [43] is contained in the following theorem.

**Theorem 4.7.** ([43, Theorem 1]) *If* (4.22) *holds and*

$$\int_0^\infty u^{2n-1} r^2(u)du < \infty, \tag{4.31}$$

*then no nontrivial solution of equation* (4.30) *satisfies*

$$\int_0^\infty f^2(y(u))du < \infty. \tag{4.32}$$

If $n = 2$ in Theorem 4.7, then we obtain Detki's result, Theorem 4.4 above. It also agrees with Theorems 4.5 and 4.6 as well. If, in Theorem 4.7, the function $f$ is linear, then we have the following corollary.

**Corollary 4.2.** *If* (4.31) *holds, then every nontrivial solution of*

$$y^{(n)} + r(t)y = 0 \tag{4.33}$$

*satisfies*

$$\int_0^\infty y^2(u)du = \infty,$$

*i.e., no nontrivial solution of* (4.33) *belongs to* $L^2[0, \infty)$.

With $n = 2$, Corollary 4.2 is Wintner's result, Theorem 4.1 above. Similar to what was done in Example 4.2, it can be shown that condition (4.31) cannot be replaced by (4.28).

**Remark 4.2.** Grammatikopoulos and Kulenović [67] obtain a result similar to Theorem 4.7 under slightly better conditions on $r$ than is required in (4.31), but with a condition on the nonlinear function $f$ that is not easily verified (see [67, p. 134–135]).

**Remark 4.3.** As Elias [43, p. 190] points out, if $a$ and $b$ are positive numbers with $1/a + 1/b = 1$ and

$$\int_0^\infty u^{2b-1}|r(u)|^b du < \infty,$$

then it can be shown that no nontrivial solution $y$ of (4.30) satisfies

$$\int_0^\infty f^a(y(u))du < \infty.$$

Detki [32, Remark 1] points out a similar result in the second order case. Also, if the function $f$ satisfies a condition like

$$f^2(u) \leq Ku^2$$

for some constant $K > 0$ and all $u$, then a conclusion like (4.23) implies that no nontrivial solution belongs to $L^2[0, \infty)$.

## 4.5. Some New Generalizations of the Early Results

In this section, we examine some more recent results obtained by Graef and Spikes [65] for nonlinear equations of the type studied in Section 4.3. Note that we *do not* require $r(t) > 0$ for the results in this section.

**Theorem 4.8.** *Assume that (4.5) holds,*

$$\int_0^\infty [1/a(u)]du = \infty, \tag{4.34}$$

*and there are constants $M_1 > 0$ and $M_2 > 0$ such that*

$$|f(u)| > M_1 \quad \text{for all} \quad |u| > M_2. \tag{4.35}$$

*If equation (3.1) has a solution $y$ satisfying*

$$\int_0^\infty f^2(y(u))du < \infty, \tag{4.36}$$

*then $a(t)y'(t) \to 0$ as $t \to \infty$ and*

$$\int_0^\infty a(u)|y'(u)|du < \infty.$$

*Proof.* Let $y$ be a nontrivial solution of equation (3.1) satisfying (4.36) and assume that $a(t)y'(t) \not\to 0$ as $t \to \infty$. Then there exists a constant $k_1 > 0$ and an increasing sequence $\{t_n\} \to \infty$ as $n \to \infty$ such that $|a(t_n)y'(t_n)| > k_1$ for $n \geq 1$. By Schwarz's inequality,

$$\int_0^\infty |r(u)||f(y(u))|\,du \leq \left(\int_0^\infty r^2(u)\,du\right)^{\frac{1}{2}} \left(\int_0^\infty f^2(y(u))\,du\right)^{\frac{1}{2}} < \infty.$$

Thus, there is a positive integer $N$ such that

$$\int_{t_N}^\infty |r(u)f(y(u))|\,du < k_1/2.$$

Integrating equation (3.1), we obtain

$$a(t)y'(t) = a(t_N)y'(t_N) - \int_{t_N}^t r(u)f(y(u))\,du,$$

so

$$|a(t)y'(t)| \geq |a(t_N)y'(t_N)| - \int_{t_N}^t |r(u)f(y(u))|\,du > k_1/2$$

for $t \geq t_N$. Hence, either

$$y'(t) > k_1/2a(t) \quad \text{or} \quad y'(t) < -k_1/2a(t)$$

for $t \geq t_N$. In either case, integrating and applying (4.34) shows that $|y(t)| \to \infty$ as $t \to \infty$. In view of (4.35), this contradicts (4.36), so $a(t)y'(t) \to 0$ as $t \to \infty$.

Next, we integrate (3.1) from $t$ to $s$, let $s \to \infty$, and integrate again, to obtain

$$\int_0^t a(u)|y'(u)|\,du \leq \int_0^t \left(\int_u^\infty |r(w)f(y(w))|\,dw\right)du.$$

Integrating by parts, we have

$$\int_0^t \left(\int_u^\infty |r(w)f(y(w))|\,dw\right)du = t\int_t^\infty |r(u)f(y(u))|\,du$$
$$+ \int_0^t u|r(u)f(y(u))|\,du \leq \int_0^\infty u|r(u)||f(y(u))|\,du.$$

Now Schwarz's inequality and (4.5) imply

$$\int_0^\infty u|r(u)||f(y(u))|\,du \leq \left(\int_0^\infty u^2 r^2(u)\,du\right)^{\frac{1}{2}} \left(\int_0^\infty f^2(y(u))\,du\right)^{\frac{1}{2}} < \infty,$$

so

$$\int_0^\infty a(u)|y'(u)|\,du < \infty,$$

and this completes the proof of the theorem.                                                    □

**Remark 4.4.** Theorem 4.8 generalizes Wintner's lemma, Lemma 4.1 above, to nonlinear equations.

The following theorem is the main result in this section.

**Theorem 4.9.** *Assume that*

$$f(u) \text{ is bounded away from zero if } u \text{ is bounded away from zero,} \qquad (4.37)$$

*(4.2) and (4.34) hold, $a(t) \geq a_1 > 0$, and there are constants $M_3 > 0$ and $M_4 > 0$ such that*

$$|f(u)| \leq M_3|u| \quad \text{if} \quad |u| < M_4. \qquad (4.38)$$

*Then no nontrivial solution of equation (3.1) satisfies (4.36), i.e., every nontrivial solution of equation (3.1) satisfies (4.23).*

*Proof.* Let $y$ be a nontrivial solution of equation (3.1). If (4.36) holds, then

$$\int_0^\infty |y'(u)|du < \infty$$

by Theorem 4.8. Hence,

$$y(t) = y(0) + \int_0^t y'(u)du \to k_2$$

as $t \to \infty$, where $k_2$ is a constant. In view of (4.36) and (4.37), we must have $k_2 = 0$. From the proof of Theorem 4.8,

$$a(t)y'(t) = \int_t^\infty r(u)f(y(u))du,$$

so dividing by $a(t)$, integrating again from $t$ to $u$, and letting $u \to \infty$, we obtain

$$y(t) = -\int_t^\infty [1/a(u)] \left( \int_u^\infty r(w)f(y(w))dw \right) du.$$

Thus,

$$|y(t)| \leq \int_t^\infty \left( \int_u^\infty |r(w)f(y(w))|dw \right) du/a_1.$$

Since $y(t) \to 0$ as $t \to \infty$, there exists $T \geq 0$ such that $|f(y(t))| \leq M_3|y(t)|$ for $t \geq T$ by (4.38). Hence,

$$|y(t)| \leq (M_3/a_1) \int_t^\infty \left( \int_u^\infty |r(w)|dw \right) du$$

for $t \geq T$. An argument similar to the one used in the proof of Theorem 4.8 shows that

$$\int_t^\infty \left( \int_u^\infty |r(w)y(w)|dw \right) du \leq \int_t^\infty u|r(u)||y(u)|du,$$

and so

$$|y(t)| \leq (M_3/a_1) \int_t^\infty u|r(u)||y(u)|du$$

for $t \geq T$. Squaring both sides of the above inequality and integrating, we have

$$\int_t^\infty y^2(u)du \leq (M_3/a_1)^2 \int_t^\infty \left( \int_u^\infty w|r(w)||y(w)|dw \right)^2 du$$

for $t \geq T$. An application of Schwarz's inequality yields

$$\left( \int_u^\infty w|r(w)||y(w)|dw \right)^2 \leq \left( \int_u^\infty y^2(w)dw \right) \left( \int_u^\infty w^2 r^2(w)dw \right)$$

$$\leq \left( \int_t^\infty y^2(u)du \right) \left( \int_u^\infty w^2 r^2(w)dw \right)$$

and so

$$\int_t^\infty y^2(u)du \leq \left( (M_3/a_1)^2 \int_t^\infty y^2(u)du \right) \left( \int_t^\infty \left( \int_u^\infty w^2 r^2(w)dw \right) du \right)$$

$$\leq \left( (M_3/a_1)^2 \int_t^\infty y^2(u)du \right) \left( \int_t^\infty u^3 r^2(u)du \right)$$

for $t \geq T$. This implies

$$(M_3/a_1)^2 \int_t^\infty u^3 r^2(u)du \geq 1,$$

which contradicts (4.2) and completes the proof of the theorem.      $\square$

**Remark 4.5.** Theorem 4.9 includes Wintner's result (Theorem 4.1 above), Suyemoto and Waltman's theorem (Theorem 4.3 above), Detki's result (Theorem 4.4 above), and Wong's theorem (Theorem 4.5 above) as special cases.

**Remark 4.6.** Theorems 4.8 and 4.9 were actually proved for perturbed equations with quite general perturbation terms that permit forcing terms as a special case. For the details, we refer the reader to Theorems 10 and 11 in [65].

**Open Problems.**

**Problem 4.1.** *Consider equation*

$$(a(t)y^{(n)})^{(n)} + r(t)y^{2k-1} = 0, \tag{4.39}$$

*where $k > 1$ is a positive integer. This equation of order $2n$ is the higher order analog of the equation*

$$(a(t)y')' + r(t)y^{2k-1} = 0$$

*studied in Chapter 3. For this equation, what would be the form of the transformation analogous to (3.5).*

**Problem 4.2.** *Apply the results from problem 4.1 to equation (4.39). Any results obtained proceeding in this manner would be both very interesting and new.*

# Chapter 5

# Relationship to Other Asymptotic Properties

In this chapter, we discuss some relationships between the limit–circle property and the boundedness, oscillation, and convergence to zero of solutions of second order equations.

## 5.1. Second Order Linear Equations

An interesting topic for research is the connection between the limit–circle property and other asymptotic properties of solutions such as the boundedness, oscillation, and convergence to zero. For example, we have the following results of Patula and Waltman [98, Theorems 1 and 2].

**Theorem 5.1.** *If equation* (1.7) *is of the limit–circle type and*

$$\int_0^\infty [1/a^{\frac{1}{2}}(t)]dt = \infty, \tag{5.1}$$

*then equation* (1.7) *is oscillatory.*

**Theorem 5.2.** *If equation* (1.7) *is of the limit–circle type and there is a constant* $B_1 > 0$ *such that* $a(t) \le B_1$, *then* (1.7) *is oscillatory and the distance between consecutive zeros of any solution of* (1.7) *tends to zero as* $t \to \infty$.

The contrapositive to the statement of Theorem 5.1, namely,

> *If condition* (5.1) *holds and equation* (1.7) *is nonoscillatory, then equation* (1.7) *is of the limit–point type.*

seems somewhat surprising. This is in fact the form of this result that was proved by Patula and Wong [99, Lemma 2.1] for equation (1.14).

In the case of all these types of interrelationships, an interesting question to raise is: "limit–circle plus what other conditions will yield one of these properties." For example, Wong [119, Conjecture I] conjectured that if equation (1.14) is of the limit–circle type, then all its solutions must be bounded. Kwong [89] presents the following counterexample which he attributes, without a reference, to J. Walter.

**Example 5.1.** The equation

$$y'' + \frac{1 - g^3(t)g''(t)}{g^4(t)} y = 0 \tag{5.2}$$

has the solution

$$y(t) = g(t) \cos \left( \int_0^t \frac{1}{g^2(u)} du \right).$$

Thus, if $g \in C^2(\mathbb{R}_+) \cap L^2[0, \infty)$ is a suitably chosen positive and unbounded function, then $y$ will be an unbounded $L^2$ solution of (5.2).

Burton and Patula [24] proved the following result on the connection between boundedness and the limit–circle property.

**Theorem 5.3.** ([24, Theorem 4]) *If equation* (1.14) *is limit–circle and there are constants* $B_2, B_3 > 0$ *such that*

$$r'(t)_+/r(t) \le B_2$$

*and*

$$r(u)/r(v) \le B_3 \quad \text{for all } v \text{ and any } u \le v,$$

*then all solutions of* (1.14) *are bounded.*

Krall [85] also studied the connection between boundedness and the limit–circle property of solutions of the linear equation (1.7) and he proved the following results.

**Theorem 5.4.** ([85, Corollary 2]) *If there are positive constants* $m_1$, $m_2$, *and* $m_3$ *such that*

$$a(t) \ge m_1, \quad a(t)r(t) \ge m_2, \quad |(a(t)r(t))'| \le m_3,$$

*and equation* (1.7) *is of the limit–circle type, then all solutions of* (1.7) *are bounded.*

**Theorem 5.5.** ([85, Corollary 4]) *If there are positive constants $m_4$, $m_5$, $m_6$, and $m_7$ such that*

$$a(t) \geq m_4, \quad |a'(t)| \leq m_5, \quad r(t) \geq m_6, \quad |r'(t)| \leq m_7,$$

*and equation* (1.7) *is of the limit–circle type, then all solutions of* (1.7) *are bounded.*

The following theorem indicates one possible relationship between the limit–circle property and the convergence to zero of solutions of (1.7). It was also proved by Burton and Patula [24].

**Theorem 5.6.** ([24, Theorem 5]) *Suppose that equation* (1.14) *is limit–circle and there are positive constants $B_4$ and $B_5$ such that*

$$\left| \frac{r'(t)}{r^{\frac{3}{2}}(t)} \right| \leq B_4$$

*and*

$$\frac{-r'(t)}{r(t)} \leq B_5.$$

*Then every solution of* (1.14) *converges to zero as $t \to \infty$.*

Other results on the convergence to zero of solutions of limit–circle type equations of the form (1.14) can be found in [24].

The following example shows that there do exist equations of the limit–circle type having an oscillatory coefficient function $r$.

**Example 5.2.** Consider equation (2.1) with

$$r(t) = \alpha^2(t) + \frac{\alpha''(t)}{2\alpha(t)} - \frac{3}{4} \frac{[\alpha'(t)]^2}{\alpha^2(t)}, \quad t \in \mathbb{R}_+$$

where

$$\alpha(t) = t^2 + (t - k + 1)^4 (k + 1 - t)^4 \exp\{(3 - 2(k + (k^2 + 1)^2)^2)$$
$$\times (t - k)^2 + 2(k + (k^2 + 1)^2)(t - k)\}$$
$$\text{for} \quad t \in \mathbb{R}_+ \cap (k - 1, k + 1], \quad k = 0, 2, 4, \ldots.$$

It is easy to see that $\alpha \in C^2(\mathbb{R}_+)$ and $\alpha \geq t^2$ on $\mathbb{R}_+$. By Theorem 1.12, this equation is of the limit–circle type. Since

$$r(k) = k^4 > 0 \quad \text{for} \quad k = 1, 3, 5, \ldots$$

and

$$r(k) = \frac{1}{(k^2 + 1)^2} [-2(k^2 + 1)^4 - 12k^2 - 12k(k^2 + 1)] < 0 \quad \text{for} \quad k = 2, 4, \ldots,$$

$r$ is oscillatory.

## 5.2. Second Order Nonlinear Equations

Next, we discuss relationships between the nonlinear limit–circle property and the boundedness, oscillation, and convergence to zero of solutions of the nonlinear equations

$$(a(t)y')' + r(t)y^{2k-1} = 0 \tag{5.3}$$

and

$$(a(t)y')' + r(t)|y|^\gamma \operatorname{sgn} y = 0, \quad 0 < \gamma \le 1 \tag{5.4}$$

similar to what we did for the linear equation (1.7) in Section 5.1.

### 5.2.1. The Superlinear Case

We begin with some consequences of the integrability criteria given in Theorem 3.1. We study the properties of nonlinear limit–circle type solutions of equation (5.3), i.e., other properties of solutions that are known to satisfy

$$\int_0^\infty y^{2k}(u)du < \infty. \tag{5.5}$$

Our first theorem relates the nonlinear limit–circle property to the oscillation of solutions. Condition (4.34) is not as good as condition (5.1); clearly, an improved result would be desirable here.

**Theorem 5.7.** *If equation (5.3) is of the nonlinear limit–circle type, and (4.34) holds, then all solutions of (5.3) are oscillatory.*

*Proof.* Let $y(t)$ be a nonoscillatory solution of (5.3), say $y(t) > 0$ for $t \ge t_1 \ge 0$. The proof in case $y(t) < 0$ for $t \ge t_1$ is similar and will be omitted. Assume that there exists $t_2 \ge t_1$ such that $y'(t_2) < 0$. Then from equation (5.3) we have $(a(t)y'(t))' \le 0$, and integrating twice we obtain

$$y(t) \le y(t_2) + \int_{t_2}^t [a(t_2)y'(t_2)/a(u)]du \to -\infty$$

as $t \to \infty$. This contradicts the fact that $y(t) > 0$ for $t \ge t_1$. Thus, $y'(t) \ge 0$ for $t \ge t_1$ which contradicts the fact that $y(t)$ is a nonlinear limit–circle solution of equation (5.3). $\qquad\square$

   If $k = 1$ so that we are considering the linear equation (1.7), then Theorem 5.7 reduces to Theorem 2(b) of Wong and Zettl [121].
   Wong and Zettl [121, p. 303] discuss the equation

$$(t^\alpha y')' + kt^{\alpha-2}y = 0, \tag{5.6}$$

where $\alpha$ and $k$ are constants, and point out that this equation is nonoscillatory if $4k \leq (\alpha - 1)^2$. In addition, equation (5.6) is of the limit–circle type if either (i) $4k > (\alpha - 1)^2$ and $\alpha > 2$, or (ii) $4k \leq (\alpha - 1)^2$ and $k > (\alpha - 3/2)/2$. In this latter case, we then have that (5.6) is nonoscillatory and of the limit–circle type as well. This shows that a condition like (5.1) or (4.34) is needed to ensure that the limit–circle property implies oscillation. Observe that for equation (5.6), condition (4.34) holds if $\alpha \leq 1$ while (5.1) holds if $\alpha \leq 2$. It is also interesting to notice that if Wong and Zettl's result in Theorem 1.11 above is applied to equation (5.6), then we see that (5.6) is of the limit–point type if $k \leq (\alpha - 3/2)/2$.

As another example, consider the equation

$$(e^{\sigma} y')' + e^{\sigma} y = 0 \tag{5.7}$$

where $\sigma$ is a constant. It is easy to see that if $|\sigma| \geq 2$, then equation (5.7) is nonoscillatory, and if $|\sigma| < 2$, it is oscillatory. In addition, if $\sigma > 0$, equation (5.7) is of the limit–circle type, and if $\sigma \leq 0$, it is of the limit–point type. Clearly, without some additional assumptions, we can draw no conclusion about the oscillation of solutions of a limit–circle type equation. Note that condition (4.34) holds if $\sigma \leq 0$ and it fails to hold if $\sigma > 0$. The same is true for condition (5.1).

The next three theorems relate the nonlinear limit–circle property to the convergence to zero of solutions of (5.3).

**Theorem 5.8.** *If*

$$\int_0^{\infty} [(a(s)r(s))'_-/a(s)r(s)]ds < \infty, \tag{5.8}$$

*then every solution of equation (5.3) is bounded. If $y(t)$ is a nonoscillatory limit–circle type solution of equation (5.3), then $y(t) \to 0$ as $t \to \infty$.*

*Proof.* We will write equation (5.3) as the system (3.35). Define $V(t)$ by

$$V(t) = a(t)w^2(t)/2r(t) + y^{2k}(t)/2k;$$

then,

$$V'(t) \leq [(a(t)r(t))'_-/a(t)r(t)]V(t).$$

An application of Gronwall's inequality and condition (5.8) shows that $V(t)$ is bounded, so $y(t)$ is bounded.

Suppose $y(t)$ is a nonoscillatory solution of (5.3), say $y(t) > 0$ for $t \geq t_1 \geq 0$. Clearly, $\liminf_{t \to \infty} y(t) = 0$. In addition, $(a(t)y'(t))' \leq 0$, so $y(t)$ is eventually monotonic, and we are done. The proof in case $y(t) < 0$ for $t \geq t_1$ is similar. $\square$

**Remark 5.1.** If $y(t)$ is a solution of (5.3) with arbitrarily large zeros but is ultimately nonnegative or nonpositive (a Z-type solution as defined in [61, 62, 63]), the argument used in proof of Theorem 5.8 shows that such solutions also converge to zero.

**Remark 5.2.** Note that the first part of Theorem 5.8 is the analogue of the boundedness result for sublinear equations given in Theorem 3.7.

**Theorem 5.9.** *Under the hypotheses of Theorem 3.1, if $a(t)r(t) \to \infty$ as $t \to \infty$, then all solutions of equation (5.3) converge to zero as $t \to \infty$.*

*Proof.* From the proof of Theorem 3.1, we have

$$(a(t)r(t))^{\beta-\alpha} y^{2k}(t) \leq M_2,$$

so

$$y^{2k}(t) \leq M_2/(a(t)r(t))^{\beta-\alpha} \to 0$$

as $t \to \infty$. $\square$

**Theorem 5.10.** *Suppose conditions (3.28), (3.29), and (5.8) hold. If $y(t)$ is a nonlinear limit–circle type solution of equation (5.3), then $y(t) \to 0$ as $t \to \infty$.*

*Proof.* Define $V(t)$ as in the proof of Theorem 5.8. By Theorem 3.2 and the fact that $x(t)$ is a nonlinear limit–circle type solution of (5.3), we have

$$\int_0^\infty V(u)du < \infty.$$

Hence, there exists an increasing sequence $\{t_n\} \to \infty$ as $n \to \infty$ such that $V(t_n) \to 0$ as $n \to \infty$. Let $\varepsilon > 0$ be given and choose $N_\varepsilon$ such that $V(t_n) < \varepsilon/4$ for $n > N_\varepsilon$. Since (5.8) holds, choose $m > N_\varepsilon$ such that

$$\exp \int_{t_m}^\infty [(a(u)r(u))'_-/a(u)r(u)]du < 2.$$

Integrating $V'(t)$, we obtain

$$V(t) \leq V(t_m) + \int_{t_m}^t [(a(u)r(u))'_-/a(u)r(u)]V(u)du$$

and by Gronwall's inequality

$$V(t) \leq (\varepsilon/4) \int_{t_m}^t [(a(u)r(u))'_-/a(u)r(u)]du < \varepsilon.$$

Hence, $V(t) \to 0$ as $t \to \infty$ and so $y(t) \to 0$ as $t \to \infty$. $\square$

**Remark 5.3.** In [105, Theorem 1], Spikes gives some sufficient conditions for all solutions of equation (3.1) to be of the nonlinear limit–point type, i.e., to satisfy (3.2). In [105, Corollary 3], he modifies the hypotheses slightly and shows that all solutions must converge to zero as $t \to \infty$. For additional results on nonlinear limit–circle solutions converging to zero, see Theorems 3, 4, and 6–8 of Graef and Spikes [65].

### 5.2.2. The Sublinear Case

First, observe that $y(t)$ being a nonlinear limit–circle solution of (5.4) means (3.55) holds. Clearly, the oscillation result Theorem 5.7 for superlinear equations also holds for the sublinear equation (5.4) as well. A boundedness result for sublinear equations was given in Theorem 3.7 above.

The next three theorems relate the nonlinear limit–circle property to convergence to zero of solutions of (5.4).

**Theorem 5.11.** *If $y(t)$ is a nonoscillatory limit–circle type solution of equation (5.4), then $y(t) \to 0$ as $t \to \infty$.*

*Proof.* The proof is the same as the second part of the proof of Theorem 5.8 and will not be given. $\qquad\square$

**Theorem 5.12.** *If (3.59)–(3.60) hold and $a(t)r(t) \to \infty$ as $t \to \infty$, then all solutions of equation (5.4) converge to zero as $t \to \infty$.*

*Proof.* From the proof of Theorem 3.8, we have

$$(a(t)r(t))^{k/(k+1)}y^{\gamma+1}(t) \le K_2(\gamma + 1),$$

so

$$y^{\gamma+1}(t) \le K_2(\gamma + 1)/(a(t)r(t))^{k/(k+1)} \to 0$$

as $t \to \infty$. $\qquad\square$

**Theorem 5.13.** *Suppose that (3.29) and (3.51) hold. If $y(t)$ is a nonlinear limit–circle type solution of (5.4), then $y(t) \to 0$ as $t \to \infty$.*

*Proof.* Define $V(t)$ as in the proof of Theorem 3.7, i.e.,

$$V(t) = a(t)w^2(t)/2r(t) + y^{\gamma+1}(t)/(\gamma + 1).$$

If $y(t)$ is a nonlinear limit–circle type solution, then Theorem 3.9 implies

$$\int_0^\infty V(u)\,du < \infty.$$

Hence, there exists a sequence $\{t_n\} \to \infty$ as $n \to \infty$ such that $V(t_n) \to 0$ as $n \to \infty$. Let $\varepsilon > 0$ be given and choose $N_\varepsilon$ so that $V(t_n) < \varepsilon/4$ for $n > N_\varepsilon$. Since (3.51) holds, choose $m > N_\varepsilon$ such that

$$\exp \int_{t_m}^{\infty} [(a(u)r(u))'_-/a(u)r(u)]du < 2.$$

Integrating $V'(t)$, we obtain

$$V(t) \leq V(t_m) + \int_{t_m}^{t} [(a(u)r(u))'_-/a(u)r(u)]V(u)du.$$

Gronwall's inequality implies

$$V(t) \leq (\varepsilon/4) \exp \int_{t_m}^{t} [(a(u)r(u))'_-/a(u)r(u)]du < \varepsilon.$$

Hence, $V(t) \to 0$ as $t \to \infty$ and so $y(t) \to 0$ as $t \to \infty$.                $\square$

**Remark 5.4.** Many of the above results can be extended to the equation (3.47) (see Graef [60]).

**Open Problems.**

**Problem 5.1.** *Prove Theorem 5.7 with condition (4.34) replaced by (5.1), or give a counterexample to show that such a result does not hold.*

**Problem 5.2.** *Burton and Patula [24, Corollary 1] show that if equation (1.14) is of the limit circle–type and $|r'(t)/r^{\frac{3}{2}}(t)| < \infty$, then*

$$\int_{0}^{\infty} \frac{1}{r^{\frac{1}{2}}(u)}du < \infty.$$

*Prove that such a result holds for nonlinear equations or give a counterexample.*

**Problem 5.3.** *Assuming that equation (1.7) or (1.14) is of the limit–circle type, what additional conditions are needed to ensure that*

$$\int_{0}^{\infty} \frac{1}{r^{\frac{1}{2}}(u)}du < \infty.$$

**Problem 5.4.** *Same as Problem 5.3 above except do it for the nonlinear equations (5.3), (5.4), or the more general equation (3.1).*

**Problem 5.5.** *Similar to Problems 5.3 and 5.4 above, assume that the equation is limit–point and determine what conditions can be added in order to ensure that*

$$\int_0^\infty \frac{1}{r^{\frac{1}{2}}(u)} du = \infty.$$

**Problem 5.6.** *Pavljuk [100] shows that if*

$$\int_0^\infty \left| \frac{r''(u)}{r(u)} - \frac{5}{4} \left( \frac{r'(u)}{r(u)} \right)^2 \right| du < \infty,$$

*then $r(t) \geq r_0 > 0$ is a necessary and sufficient condition for all solutions of equation (1.14) to be bounded (also see condition (1.17) above). Extend this result to nonlinear equations.*

**Notes.** Theorems 5.5–5.7 in Section 5.2.1 are due to Graef [59] specialized to equation (5.3). Example 5.2 and the results in Section 5.2.2 for the sublinear case are new.

# Chapter 6

# Third Order Differential Equations

This chapter contains a discussion of the limit–point/limit–circle problem for third order equations. We begin with nonlinear equations with quasiderivatives and then consider some special third order linear and nonlinear equations.

## 6.1. Equations with Quasiderivatives

We consider the nonlinear third-order differential equation

$$\left(\frac{1}{a_2(t)}\left(\frac{1}{a_1(t)}y'\right)'\right)' = r(t)f(y, y^{[1]}, y^{[2]}), \qquad (6.1)$$

where $y^{[i]}$ is the $i$-th quasiderivative of $y$ (see Section 2.1), the functions $a_1, a_2 \in C^0(\mathbb{R}_+)$ are positive, $f : \mathbb{R}^3 \to \mathbb{R}$ is continuous, and $r \in L_{\text{loc}}(\mathbb{R}_+)$. Throughout this section, we assume that

$$(-1)^\alpha r(t) > 0 \quad \text{and} \quad x_1 f(x_1, x_2, x_3) \geq 0 \quad \text{on} \quad \mathbb{R}^3, \qquad (6.2)$$

where $\alpha \in \{0, 1\}$ and

$$(-1)^{\alpha+1}\left(\frac{a_1(t)}{a_2(t)}\right)' \geq 0. \qquad (6.3)$$

As in Chapter 3, we are only concerned with solutions of equation (6.1) that are continuable and not eventually identically zero. Such a solution is *oscillatory* if it has arbitrarily large zeros.

In accordance with Definition 2.1, a continuable solution $y$ of equation (6.1) is said to be of the *nonlinear limit–circle type* if it satisfies

$$\int_0^\infty y(t) f(y^{[0]}(t), y^{[1]}(t), y^{[2]}(t)) \, dt < \infty. \tag{6.4}$$

Otherwise, it is said to be of the *nonlinear limit–point type*, i.e.,

$$\int_0^\infty y(t) f(y^{[0]}(t), y^{[1]}(t), y^{[2]}(t)) \, dt = \infty. \tag{6.5}$$

The equation (6.1) is said to be of the *nonlinear limit–circle type* if all its solutions satisfy (6.4). Otherwise, the equation (6.1) is said to be of the *nonlinear limit–point type*, i.e., there is at least one solution satisfying (6.5).

Nonoscillatory solutions of equation (6.1) that are not eventually constant can be divided into the following classes (see, for example, [3]–[9]):

$$\mathbb{M}_0 = \{y : \exists t_y : y(t)y^{[1]}(t) \le 0, \ y(t)y^{[2]}(t) < 0 \text{ for } t \ge t_y\},$$
$$\mathbb{M}_1 = \{y : \exists t_y : y(t)y^{[1]}(t) \ge 0, \ y^{[1]}(t)y^{[2]}(t) < 0 \text{ for } t \ge t_y\},$$
$$\mathbb{M}_2 = \{y : \exists t_y : y(t)y^{[1]}(t) > 0, \ y(t)y^{[2]}(t) \ge 0 \quad \text{for } t \ge t_y\},$$
$$\mathbb{K} = \{y : \exists t_y : y(t)y^{[1]}(t) < 0, \ y(t)y^{[2]}(t) \ge 0 \quad \text{for } t \ge t_y\}.$$

Observe that the sign of the third quasiderivative follows from Equation (6.1) and assumption (6.2). If $\alpha = 1$ and $y$ is of class $\mathbb{K}$, then $y$ is called a *Kneser* solution (see, for example, [10]).

We begin with a limit–point result.

**Theorem 6.1.** *Let $\alpha = 0$ and suppose there exist $K > 0$ and $C > 0$ such that*

$$\frac{1}{|x_1|} \le |f(x_1, x_2, x_3)| \le C|x_1| \quad \text{for} \quad |x_1| \ge K > 0. \tag{6.6}$$

*Then equation (6.1) is of the nonlinear limit–point type.*

*Proof.* By Corollary 2.1, the right-hand inequality in (6.6) guarantees that all solutions of (6.1) are continuable. Let $y$ be a solution of (6.1) with initial conditions

$$y(0) \ge K, \ y^{[1]}(0) > 0, \ y^{[2]}(0) > 0. \tag{6.7}$$

From (6.1) and (6.2), it follows that $y$ and $y^{[1]}$ are increasing. Then, in view of (6.7), we have $y(t) \ge K$ for $t \in \mathbb{R}_+$. Thus, (6.6) implies

$$\int_0^\infty y(t) f(y(t), y^{[1]}(t), y^{[2]}(t)) \, dt \ge \int_0^\infty dt = \infty.$$

Hence, (6.1) is of the nonlinear limit–point type. $\qquad\square$

In what follows, an important role will be played by the auxiliary function

$$F(t) = (-1)^{\alpha} y(t) y^{[2]}(t) + (-1)^{\alpha+1} \frac{1}{2} \frac{a_1(t)}{a_2(t)} \left( y^{[1]}(t) \right)^2 \qquad (6.8)$$

where $y$ is a solution of (6.1).

**Lemma 6.1.** *If (6.3) holds, then*

$$F'(t) = (-1)^{\alpha} y(t) y^{[3]}(t) + (-1)^{\alpha+1} \frac{1}{2} \left( \frac{a_1(t)}{a_2(t)} \right)' \left( y^{[1]}(t) \right)^2 \geq 0. \qquad (6.9)$$

*Proof.* From (6.2) and (6.3), we have

$$F'(t) = (-1)^{\alpha} y y^{[3]} + (-1)^{\alpha} a_1 y^{[1]} y^{[2]}$$

$$+ (-1)^{\alpha+1} \frac{1}{2} \left( \frac{a_1}{a_2} \right)' (y^{[1]})^2 + (-1)^{\alpha+1} \frac{a_1}{a_2} y^{[1]} a_2 y^{[2]} \geq 0.$$

$\square$

**Theorem 6.2.** *Let $\alpha = 0$ and (6.3) hold. Suppose that $\frac{a_1(t)}{a_2(t)} \in C^1(\mathbb{R}_+)$, $r(t) \geq K > 0$, $f(x_1, x_2, x_3) \neq 0$ for $x_1 \neq 0$ in $\mathbb{R}^3$, and there exist $\beta > 1$ and $M > 0$ such that*

$$|f(x_1, x_2, x_3)| \leq \frac{1}{|x_1|^{\beta}} \qquad \text{for } |x_1| \geq M > 0, \qquad (6.10)$$

*and*

$$\int_1^{\infty} \frac{dt}{\left( \int_0^t a_1(s) \int_0^s a_2(\tau) \, d\tau \, ds \right)^{\beta-1}} < \infty. \qquad (6.11)$$

*Then equation (6.1) is of the nonlinear limit–circle type.*

*Proof.* First observe that from (6.11) and (6.3) we obtain $\int_0^{\infty} a_2(t) \, dt = \infty$. By Corollary 2.1, there are no singular solutions, i.e., every solution of (6.1) is defined on $\mathbb{R}_+$. If a solution is identically zero for large $t$, then it is obviously of the limit–circle type.

Suppose $y$ is a proper solution of (6.1). Then, by a result in [3, Theorem 6 (ii)], $y$ is either oscillatory, of class $\mathbf{M}_i$, $i = 0, 1, 2$, or $y \equiv c \neq 0$, i.e., $\mathbb{K} = \emptyset$. Moreover, it is shown there that if $y$ is oscillatory, then there exists $\tau \geq 0$ such that for any zero $\bar{t} \geq \tau$ of the function $y^{[1]}$, we have $y(\bar{t}) y^{[2]}(\bar{t}) < 0$. Let $t_0 = \max\{\tau, t_y\}$.

Let $y$ be oscillatory, or belong to the class $\mathbf{M}_i$ for $i = 0$ or 1, or let $y \equiv c \neq 0$ for large $t$. From the definition of the function $F$ and Lemma 6.1, it follows that

$$F(t) \leq 0, \quad t \in [t_0, \infty).$$

Hence, in view of (6.3), we have

$$\infty > F(\infty) - F(t_0) = \int_{t_0}^{\infty} F'(s)\, ds \geq \int_{t_0}^{\infty} y(t)y^{[3]}(t)\, dt$$

$$= \int_{t_0}^{\infty} y(t)r(t)f(y(t), y^{[1]}(t), y^{[2]}(t))\, dt$$

$$\geq K \int_{t_0}^{\infty} y(t)f(y(t), y^{[1]}(t), y^{[2]}(t))\, dt,$$

i.e., all such solutions are of the limit–circle type.

Now let $y$ belong to the class $M_2$. For definiteness, suppose $y(t) > 0$, $y^{[1]}(t) > 0$, and $y^{[2]}(t) \geq 0$ on $[t_0, \infty)$. From (6.2) and the assumptions of the theorem, it follows that

$$y^{[3]}(\tau) = 0 \quad \text{if and only if} \quad y(\tau) = 0 \quad \text{a.e..} \tag{6.12}$$

Moreover, $y$ and $y^{[1]}$ are increasing and $y^{[2]}$ is nondecreasing on $[t_0, \infty)$, so from (6.12), it is obvious that $y^{[2]} \neq 0$ on $[t_0, \infty)$, i.e.,

$$y(t) > 0, \quad y^{[1]}(t) > 0, \quad y^{[2]}(t) > 0 \quad \text{on } [t_0, \infty). \tag{6.13}$$

From (6.12) and (6.13), we obtain

$$y^{[1]}(t) \geq \int_{t_0}^{t} a_2(s)y^{[2]}(s)\, ds \geq y^{[2]}(t_0) \int_{t_0}^{t} a_2(s)\, ds$$

and

$$y(t) \geq \int_{t_0}^{t} a_1(s)y^{[1]}(s)\, ds \geq y^{[2]}(t_0) \int_{t_0}^{t} a_1(s) \int_{t_0}^{s} a_2(\tau)\, d\tau\, ds.$$

Using the inequalities

$$\int_{t_0}^{t} a_1(s) \int_{t_0}^{s} a_2(\tau)\, d\tau\, ds \geq c_1 \int_{t_0+1}^{t} a_1(s) \int_{0}^{s} a_2(\tau)\, d\tau\, ds$$

$$\geq c_2 \int_{0}^{t} a_1(s) \int_{0}^{s} a_2(\tau)\, d\tau\, ds, \quad t \geq t_0 + 2,$$

where

$$c_1 = \frac{\int_{t_0}^{t_0+1} a_2(\tau)\, d\tau}{\int_{0}^{t_0+1} a_2(\tau)\, d\tau} \quad \text{and} \quad c_2 = c_1 \frac{\int_{t_0+1}^{t_0+2} a_1(s) \int_{0}^{s} a_2(\tau)\, d\tau\, ds}{\int_{0}^{t_0+2} a_1(s) \int_{0}^{s} a_2(\tau)\, d\tau\, ds} > 0,$$

we obtain

$$y(t) \geq y^{[2]}(t_0)c_2 \int_{0}^{t} a_1(s) \int_{0}^{s} a_2(\tau)\, d\tau\, ds, \quad t \in [t_0 + 2, \infty), \tag{6.14}$$

and so (6.11) implies $\lim\limits_{t\to\infty} y(t) = \infty$. Let $t_1 \geq t_0 + 2$ be such that $y(t_1) \geq M$, where $M$ is from condition (6.10). Then, $y(t) \geq M$ for $t \in [t_1, \infty)$, and using (6.10) and (6.14), we obtain

$$\int_{t_1}^{\infty} y(t)f(y(t), y^{[1]}(t), y^{[2]}(t))\, dt \leq \int_{t_1}^{\infty} \frac{dt}{y^{\beta-1}(t)}$$

$$\leq K_1 \int_{t_1}^{\infty} \frac{dt}{\left(\int_0^t a_1(s) \int_0^s a_2(\tau)\, d\tau\, ds\right)^{\beta-1}} < \infty,$$

where $K_1 = \left(1/y^{[2]}(t_0)c_2\right)^{\beta-1} > 0$. Thus, this solution is of the nonlinear limit–circle type. $\qquad\square$

**Remark 6.1.** For the equation without quasiderivatives, i.e., $a_2 \equiv a_1 \equiv 1$, condition (6.11) is satisfied if $\beta > \frac{3}{2}$.

In order to consider the case $\alpha = 1$, we must first obtain a Kolmogorov–Horny type inequality for quasiderivatives.

**Lemma 6.2.** *Let $-\infty < t_1 < t_2 < \infty$ and let $w$, $p$, $s$, and $z$ be continuous positive functions such that the quasiderivatives $z^{[i]}$ defined by*

$$z^{[1]} = \frac{1}{w(t)}z', \quad z^{[2]} = \frac{1}{p(t)}(z^{[1]})', \quad z^{[3]} = \frac{1}{s(t)}(z^{[2]})'$$

*are continuous for $i = 1, 2$ and $z^{[3]} \in L_{loc}[t_1, t_2]$. Suppose that $z^{[i]}$, $i = 1, 2$, each have a zero in $[t_1, t_2]$ and let*

$$c_1 = \max_{[t_1, t_2]} \frac{p(t)}{w(t)}, \quad c_2 = \max_{[t_1, t_2]} \frac{s(t)}{p(t)}.$$

*Let*

$$v_i = \max_{t_1 \leq t \leq t_2} |z^{[i]}(t)|, \quad i = 0, 1, 2, \quad and \quad v_3 \geq |z^{[3]}(t)| \quad a.\,e. \text{ on } [t_1, t_2].$$

*Then,*

$$v_1 \leq K v_0^{\frac{2}{3}} v_3^{\frac{1}{3}}, \tag{6.15}$$

*where $K = 2c_1^{\frac{2}{3}} c_2^{\frac{1}{3}}$.*

*Proof.* We can choose intervals $J_i \subset [t_1, t_2]$, $i = 1, 2$ such that

$$v_i = \max_{J_i} |z^{[i]}(t)|, \quad \min_{J_i} |z^{[i]}(t)| = 0, \quad and \quad z^{[i]} \text{ does not change sign on } J_i.$$

Then,

$$v_1^2 \le 2 \int_{J_1} |z^{[1]} z^{[1]'}| \, dt \le 2v_2 \int_{J_1} \frac{p}{r} |z'| \le 2v_2 c_1 \int_{J_1} |z'| \, dt \le 2c_1 v_0 v_2$$

and

$$v_2^2 \le 2 \int_{J_2} s(t) |z^{[2]} z^{[3]}| \, dt \le 2v_3 c_2 \int_{J_1} z^{[1]'} \, dt \le 2c_2 v_1 v_3.$$

Thus,

$$v_1^2 \le 2c_1 v_0 v_2 \le 2c_1 v_0 \sqrt{2c_2 v_1 v_3},$$

i.e., $v_1 \le K v_0^{\frac{2}{3}} v_3^{\frac{1}{3}}$.                                                                    □

The following two theorems apply to the case where $\alpha = 1$.

**Theorem 6.3.** *Let $\alpha = 1$ and (6.3) hold. Suppose that there exist $K > 0$ and $L_i > 0$, $i = 1, 2$, such that (6.6) holds,*

$$|f(x_1, x_2, x_3)| \le 1 + |x_1| \quad on \ \mathbb{R}^3, \tag{6.16}$$

$$\frac{|r(t)|}{a_2(t)} \le L_1, \quad \frac{a_1(t)}{a_2(t)} \le L_2, \tag{6.17}$$

$$\int_0^\infty a_1(t) \, dt = \infty, \quad and \ a_1 \ is \ bounded \ from \ above. \tag{6.18}$$

*Then equation (6.1) is of the nonlinear limit–point type.*

*Proof.* From (6.16) and Corollary 2.1, it follows that every solution of (6.1) is defined on $\mathbb{R}_+$. Under hypotheses (6.17) and (6.18), every nonoscillatory solution is either of Kneser type, of class $M_2$, or identically zero for large $t$.

From (6.16) and Corollary 2.1, it follows that the limit of the function $F$ in (6.8) satisfies

$$-\infty < F(\infty) \le 0$$

for a Kneser solution. Since $F$ is nondecreasing, for any solution of (6.1) with initial conditions satisfying $F(0) > 0$, this solution will be either oscillatory or of class $M_2$. Let $y$ be a solution of (6.1) satisfying initial conditions such that

$$F(0) \ge 2(1 + 2K)^2 L_1^{2/3} L_2 \left( \frac{a_2(0)}{a_1(0)} \right)^{4/3} \tag{6.19}$$

Suppose $y$ is oscillatory. Clearly, $y$ is also a solution of

$$\frac{1}{a_2(t)} \left( \frac{1}{a_2(t)} \left( \frac{1}{a_1(t)} y' \right)' \right)' = \frac{r(t)}{a_2(t)} f(y, y^{[1]}, y^{[2]}).$$

From [8, Theorem 2.1], it follows that there exist increasing sequences $\{t_k^i\}$, $i = 0, 1, 2$, and $\{\bar{t}_k^2\}$, tending to $\infty$ as $k \to \infty$, such that

$$t_k^0 < t_k^2 \le \bar{t}_k^2 < t_k^1 < t_{k+1}^0,$$

$$y^{[i]}(t_k^i) = 0, \ i = 0, 1, 2, \qquad y^{[2]}(t) \equiv 0 \quad \text{for} \quad t \in [t_k^2, \bar{t}_k^2],$$

$$yy^{[1]} \begin{cases} > 0 & \text{on } (t_k^0, t_k^1), \\ < 0 & \text{on } (t_k^1, t_{k+1}^0), \end{cases} \qquad yy^{[2]} \begin{cases} > 0 & \text{on } (t_k^0, t_k^2), \\ < 0 & \text{on } (\bar{t}_k^2, t_{k+1}^0). \end{cases} \tag{6.20}$$

Let

$$v_{ik} = \max_{0 \le t \le t_k^1} |y^{[i]}(t)|, \ i = 0, 1, 2, \ \text{and} \ v_{3k} = L_1(1 + v_{0k}).$$

Under assumption (6.3), the Kolmogorov–Horny type inequality (Lemma 6.2)

$$v_{1k} \le 2\left(\frac{a_2(0)}{a_1(0)}\right)^{2/3} v_{0k}^{\frac{2}{3}} v_{3k}^{\frac{1}{3}} \le M_1(1 + v_{0k})$$

holds, where $M_1 = 2\left(\frac{a_2(0)}{a_1(0)}\right)^{2/3} L_1^{1/3}$. From definition of the function $F$, we have

$$F(t_k^0) = \frac{1}{2}\frac{a_1}{a_2} y^{[1]2}|_{t=t_k^0}.$$

Then, for $K_2 = \sqrt{\frac{L_2}{2}}$,

$$1 + v_{0k} \ge \frac{v_{1k}}{M_1} \ge \frac{1}{M_1}|y^{[1]}(t_k^0)| \ge \frac{2}{M_1 K_2}\sqrt{\frac{a_1}{2a_2}} y^{[1]2}|_{t=t_k^0}$$

$$\ge \frac{2}{M_1 K_2}\sqrt{F(t_k^0)} \ge \frac{2}{M_1 K_2}\sqrt{F(0)}. \tag{6.21}$$

Hence, from (6.19) and (6.21) we have

$$v_{0k} \ge 2K, \quad k \in \mathbb{N}. \tag{6.22}$$

Now define the sequence of intervals $\delta_k = [s_k, t_k^1]$, $k \in \mathbb{N}$, by

$$s_k \in (t_k^0, t_k^1) \quad \text{and} \quad |y(s_k)| = \frac{v_{0k}}{2}.$$

Thus, by (6.20) and (6.22) we have $|y(t)| \ge \frac{v_{0k}}{2} \ge K$ on $\delta_k$. Let $\Delta_k = t_k^1 - s_k$. Using Lemma 6.2 and (6.22), we obtain

$$v_{0k} - \frac{v_{0k}}{2} = \frac{v_{0k}}{2} \le \int_{\delta_k} |y'(t)| \, dt \le \Delta_k v_{1k} \le \Delta_k M_1(1 + v_{0k}) \le M_1\left(1 + \frac{1}{2K}\right)\Delta_k v_{0k}.$$

Thus,

$$\Delta_k \geq c > 0,$$

for some positive constant $c$.

If $y$ is of the nonlinear limit–circle type, then using (6.6) we obtain

$$\infty > \int_0^\infty y(t) f(y(t), y^{[1]}(t), y^{[2]}(t)) \, dt$$

$$\geq \sum_{k=1}^\infty \int_{\Delta_k} y(t) f(y(t), y^{[1]}(t), y^{[2]}(t)) \, dt \geq \sum_{k=1}^\infty \int_{\Delta_k} dt = \infty,$$

which is a contradiction.

Now let $y$ be a nonoscillatory solution of class $\mathbb{M}_2$. From (6.18), we have

$$|y(t) - y(t_0)| = \int_{t_0}^t |y'(s)| \, ds = \int_{t_0}^t r(s)|y^{[1]}(s)| \, ds$$

$$\geq |y^{[1]}(t_0)| \int_{t_0}^t r(s) \, ds \to \infty \quad \text{for } t \to \infty,$$

and so $\lim_{t \to \infty} |y(t)| = \infty$. Also,

$$\int_{t_0}^\infty y(t) f(y(t), y^{[1]}(t), y^{[2]}(t)) \, dt$$

$$\geq \int_{t_1}^\infty y(t) f(y(t), y^{[1]}(t), y^{[2]}(t)) \, dt \geq \int_{t_1}^\infty dt = \infty,$$

where $t_1 \geq t_0$ is such that $|y(t_1)| \geq K$.                                     $\square$

**Theorem 6.4.** *Let $\alpha = 1$.*

*a) Suppose there exist $\varepsilon > 0$ and $K > 0$ such that $f(x_1, x_2, x_3) \neq 0$ for $x_1 \neq 0$, $|x_i| < \varepsilon$, $i = 2, 3$,*

$$|f(x_1, x_2, x_3)| \leq K|x_1| \text{ on } \mathbb{R}^3,$$

$$\int_0^\infty a_1(s) \, ds = \int_0^\infty a_2(s) \, ds = \infty,$$

*and*

$$\int_0^\infty |r(t)| \int_0^t a_2(s) \int_0^s a_1(\tau) \, d\tau \, ds \, dt < \infty. \tag{6.23}$$

*Then every Kneser solution of equation (6.1) is of the nonlinear limit–point type.*

*b) Suppose* (6.3) *holds,* $|r(t)| \geq K > 0$, *and*

$$\int_0^\infty a_1(s)\, ds = \infty.$$

*Then every Kneser solution of equation* (6.1) *is of the nonlinear limit–circle type.*

*Proof.* a) Let $y$ be any Kneser solution of (6.1). By a direct integration, it satisfies $\lim_{t \to \infty} y^{[i]} = 0$ for $i = 1, 2$. We will prove that $y$ tends to a nonzero constant $c$. Suppose, for the sake of a contradiction, that there exists a Kneser solution $y$ such that $\lim_{t \to \infty} y(t) = 0$. Without loss of generality, we assume that $y(t) > 0$ for large $t$. Successive integrations give

$$y(t) = \int_t^\infty a_1(s) \int_s^\infty a_2(u) \int_u^\infty |r(v)|\, f\big(y(v), y^{[1]}(v), y^{[2]}(v)\big)\, dv\, du\, ds.$$

From this, we have

$$y(t) \leq K y(t) \int_t^\infty a_1(s) \int_s^\infty a_2(u) \int_u^\infty |r(v)|\, dv\, du\, ds,$$

that is,

$$\frac{1}{K} \leq \int_t^\infty a_1(s) \int_s^\infty a_2(u) \int_u^\infty |r(v)|\, dv\, du\, ds.$$

Interchanging the order of integration, we obtain a contradiction to (6.23). Therefore, there exists $\tau \geq t_y$ such that $|y(t)| \geq c/2$ for $t \geq \tau$. Let us set

$$m = \min\{|f(x_1, x_2, x_3)| : c/2 \leq |x_1| \leq |y(\tau)|,\ 0 \leq |x_i| \leq |y^{[i-1]}(\tau)|,\ i = 2, 3\};$$

then $m > 0$ and

$$\int_0^\infty y(t) f(y(t), y^{[1]}(t), y^{[2]}(t))\, dt \geq \frac{cm}{2} \int_\tau^\infty dt = \infty.$$

b) As in the proof of Theorem 6.3, by [8, Theorem 3.1], $F(\infty) \leq 0$ for every Kneser solution. Integrating (6.9), we obtain

$$F(\infty) - F(0) = \int_0^\infty |r(t)|\, y(t) f(y(t), y^{[1]}(t), y^{[2]}(t))\, dt$$

$$+ \frac{1}{2} \int^\infty \left(\frac{a_1(t)}{a_2(t)}\right)' (y^{[1]}(t))^2\, dt.$$

It follows that $F(\infty) < \infty$ if and only if both integrals on the right-hand side are convergent. Since $|r|$ is bounded from below, the Kneser solution satisfies (6.4), which is what we wished to prove.                                        □

## 6.2. Linear Equations

In this section, we give limit–point/limit–circle criteria for the linear equation

$$y''' + q(t)y' + r(t)y = 0, \tag{6.24}$$

where $q \in C^1(\mathbb{R}_+), r \in C^0(\mathbb{R}_+)$ ($q$ and $r$ may change their signs). In conjunction with (6.24), we consider the second order linear equation

$$h'' + \frac{q(t)}{4}h = 0. \tag{6.25}$$

A special case of equation (6.24) is the linear equation known as *Appel's equation*,

$$y''' + q(t)y' + \frac{q'(t)}{2}y = 0. \tag{6.26}$$

For this equation we have the following result.

**Theorem 6.5.** *Let all solutions of* (6.25) *be bounded. Equation* (6.25) *is of the limit–circle type if and only if the Appel's equation is of the limit–circle type.*

*Proof.* First, recall that equation (6.26) has three linearly independent solutions of the form $u_1^2$, $u_1u_2$, $u_2^2$ where $u_i$, $i = 1, 2$, are linearly independent solutions of (6.25). Assume that equation (6.25) is of the limit–circle type. Since every solution of (6.25) belongs to $L^2$ and, by assumption, is bounded, $u$ also belongs to $L^4$. Now if $y$ is any solution of (6.26), then, for some $c_i$, $i = 1, 2, 3$,

$$\int_0^\infty y^2(t)\, dt = \int_0^\infty \left(c_1u_1^2 + c_2u_1u_2 + c_3u_2^2\right)^2 dt < \infty.$$

The converse follows from this and boundedness of solutions of (6.25). $\qquad\square$

**Remark 6.2.** If all solutions of equation (6.25) are bounded and belong to $L^2$, then equation (6.25) is oscillatory by Theorem 5.1. Hence, equation (6.26) has an oscillatory solution.

A natural question that arises is "does Theorem 6.5 hold for perturbed equations as well." The answer is given in the following result.

**Theorem 6.6.** *Let all solutions of* (6.25) *be bounded.*
   *(i) If*

$$\int_0^\infty \left| r(u) - \frac{q'(u)}{2} \right| du < \infty \tag{6.27}$$

and equation (6.25) is of the limit–circle type, then equation (6.24) is of the limit–circle type as well.

(ii) Assume that

$$\int_0^\infty \left( \int_u^\infty \left| r(s) - \frac{q'(s)}{2} \right| ds \right)^2 du < \infty. \tag{6.28}$$

Equation (6.25) is of the limit–circle type if and only if equation (6.24) is of the limit–circle type.

For the proof we need two lemmas; for simplicity, we state them for the equation

$$y'''(t) + q(t)y'(t) + \left( \frac{q'(t)}{2} + R(t) \right) y(t) = 0 \tag{6.29}$$

where $R \in C(\mathbb{R}_+)$.

**Lemma 6.3.** ([68, Theorem 3.16]) *If every solution of equation (6.25) is bounded and*

$$\int_0^\infty |R(u)| du < \infty, \tag{6.30}$$

*then every solution of equation (6.29) is bounded.*

The following lemma is a consequence of Theorem 1 in [35].

**Lemma 6.4.** *Assume that (6.30) holds and that $h_1$ and $h_2$ are bounded solutions of equation (6.25) with $h_1(t)h_2'(t) - h_1'(t)h_2(t) \equiv 1$. Let $X$ and $Y$ be the sets of all solutions of (6.26) and (6.29) on $[0, \infty)$, respectively. Then the mapping $V : Y \to X$ defined by*

$$(Vy)(t) = y(t) - \int_t^\infty K(t, s)R(s)y(s)\,ds,$$

*where*

$$K(t, s) = \begin{vmatrix} h_1(t) & h_2(t) \\ h_1(s) & h_2(s) \end{vmatrix}, \tag{6.31}$$

*is a one-to-one mapping and*

$$\lim_{t \to \infty} |x(t) - y(t)| = 0 \tag{6.32}$$

*for every $x(t) \in X$ and $y(t) \in Y$ such that $x(t) = Vy(t)$.*

*Proof of Theorem 6.6.* Equation (6.24) can be written in the form of equation (6.29) with $R(t) = r(t) - q'(t)/2$. Hence, condition (6.27) is equivalent to (6.30). Let $X$ and $Y$ be the sets of all solutions of (6.26) and (6.29) on $[0, \infty)$, respectively. By Lemma 6.3, every solution $y \in Y$ is bounded, and by Lemma 6.4, there exists a one-to-one mapping $V : Y \to X$ such that $x(t) = Vy(t)$ for $x \in X$ and $y \in Y$. Let $u(t) = x(t) - y(t)$. Then,

$$u(t) = -\int_t^\infty K(t, s)R(s)y(s)\, ds. \tag{6.33}$$

Part (i). In view of the boundedness of $h_i$, $i = 1, 2$, there exists a constant $M$ such that

$$K^2(t, s) = \left(h_1(t)h_2(s) - h_1(s)h_2(t)\right)^2$$
$$\leq 2h_1^2(t)h_2^2(s) + 2h_1^2(s)h_2^2(t) \leq 2M\left(h_1^2(t) + h_2^2(t)\right),$$

and thus

$$\int_t^\infty K(t, s)R(s)\, ds \leq \sqrt{2M}\sqrt{h_1^2(t) + h_2^2(t)}\int_t^\infty R(s)\, ds.$$

In view of (6.30), there exists a constant $N > 0$ such that $\left|\int_t^\infty R(s)\, ds\right| \leq N$ for all $t \geq 0$. From this and (6.33), it follows that

$$\int_0^\infty u^2(t)\, dt = \int_0^\infty \left(\int_t^\infty K(t, s)R(s)y(s)\, ds\right)^2 dt$$
$$\leq 2M\int_0^\infty (h_1^2(t) + h_2^2(t))\left(\int_t^\infty R(s)\, ds\right)^2 dt$$
$$\leq 2MN^2\left(\int_0^\infty h_1^2(t)\, dt + \int_0^\infty h_2^2(t)\, dt\right) < \infty$$

for every $x \in X$ and $y \in Y$ such that $x(t) = Vy(t)$. Moreover, the inequality $y^2 = (x - u)^2 \leq 2(x^2 + u^2)$ yields

$$\int_0^\infty y^2(t)\, dt \leq 2\int_0^\infty x^2(t)\, dt + 2\int_0^\infty u^2(t)\, dt. \tag{6.34}$$

By Theorem 6.5, every solution $x \in X$ is in $L^2$. Thus, every solution $y \in Y$, or equivalently, every solution of equation (6.24) belongs to $L^2$ as well.

Part (ii). First, we prove that $u \in L^2$, that is

$$\int_{t_0}^\infty (x(t) - y(t))^2\, dt < \infty \tag{6.35}$$

holds for every $x \in X$ and $y \in Y$ such that $x(t) = Vy(t)$. Clearly, (6.28) implies (6.30). From (6.33), it follows that

$$\int_0^\infty u^2(t)\, dt = \int_0^\infty \left( \int_t^\infty K(t,s) R(s) y(s)\, ds \right)^2 dt$$

$$\leq M \int_0^\infty \left( \int_t^\infty R(s)\, ds \right)^2 dt$$

for some positive constant $M$. In view of (6.28), we see that (6.35) holds.

Let equation (6.25) be of the limit–circle type. By Theorem 6.5, every solution $x \in X$ belongs to $L^2$ so let $y \in Y$ be such that $x(t) = Vy(t)$. Then, in view of (6.34), any solution $y \in Y$ is in $L^2$. Now let equation (6.24) be of the limit–circle type. Then, for any solution $x \in X$ there exists $y \in Y$ such that $x(t) = Vy(t)$. Taking into account (6.35) and using the inequality

$$x^2(t) = (y(t) + u(t))^2 \leq 2y^2(t) + 2u^2(t),$$

we have $x \in L^2$. By Theorem 6.5, equation (6.25) is of the limit–circle type. $\square$

Recall that if equation (6.25) is of the limit–circle type, then its solutions need not be bounded (see Chapter 5). However, combining Theorem 6.6 and Dunford and Schwartz's result (Theorem 1.3 above), we have the following result for equation (6.24).

**Corollary 6.1.** *Assume that $q(t) > 0$,*

$$\int_0^\infty \frac{q'_-(u)}{q(u)}\, du < \infty, \tag{6.36}$$

*and*

$$\int_0^\infty \left| \frac{q''(u)}{q^{3/2}(u)} - \frac{5}{4} \frac{[q'(u)]^2}{q^{5/2}(u)} \right| du < \infty. \tag{6.37}$$

*(i) If (6.27) and*

$$\int_0^\infty \frac{1}{q^{\frac{1}{2}}(u)}\, du < \infty \tag{6.38}$$

*hold, then equation (6.24) is of the limit–circle type.*
  *(ii) If (6.28) and*

$$\int_0^\infty \frac{1}{q^{\frac{1}{2}}(u)}\, du = \infty \tag{6.39}$$

*hold, then equation (6.24) is of the limit–point type.*

*Proof.* Observe that condition (6.36) ensures boundedness of all solutions of (6.25). By (6.37) and (6.38), every solution $h$ of (6.25) is in $L^2$. Applying Theorem 6.6(i) we obtain Part (i). By (6.37) and (6.39), there exists a solution $h$ of equation (6.25) such that $h \notin L^2$, and so Part (ii) follows from Theorem 6.6(ii).                                                                                □

**Remark 6.3.** Equations (6.26) and (6.29) satisfying (6.32) are said to be *asymptotically equivalent*. Using Lemma 6.4 and applying other asymptotic properties related to the limit–circle problem for (6.25), we can obtain further asymptotic properties of solutions of (6.29). For example, if all solutions of (6.25) converge to zero as $t \to \infty$ (for such conditions see Theorem 5.10) and (6.30) holds, then all solutions of (6.29) converge to zero as $t \to \infty$.

In the remainder of this section, we give limit–circle results for the three-term linear equation (6.24) under the assumptions $q(t) > 0$ and $r(t) > 0$ on $\mathbb{R}_+$.

Let $T \in \mathbb{R}_+$ and let $h_1$ and $h_2$ be solutions of (6.25) satisfying the Cauchy initial conditions

$$h_1(T) = 1, \ h_1'(T) = 0, \ h_2(T) = 0, \ h_2'(T) = 1.$$

Also, we set

$$h(t) = \max(|h_1(t)|, |h_2(t)|). \tag{6.40}$$

**Theorem 6.7.** *Assume (6.36) holds and that equation (6.25) is of the limit–circle type. Define the function $h$ by (6.40). If there exist $T \in \mathbb{R}_+$ and $\delta > 0$ such that*

$$\left| r(t) - \frac{q'(t)}{2} \right| \le K = \frac{1}{(2 + \delta) \int_T^\infty h^4(s) \, ds}, \quad t \in [T, \infty), \tag{6.41}$$

*then equation (6.24) is of the limit–circle type.*

For the proof of this theorem we need the following lemma which is a special case of a result of Staněk [106, Theorem 1].

**Lemma 6.5.** *Let $\varepsilon > 0$, $T \in \mathbb{R}_+$, $I = [T, \infty)$, $q \in C^1(I)$, and let $F \in C^0(I \times \mathbb{R})$ satisfy*

$$|F(t, x)| \le \omega(t, |x|) \quad \text{on } I \times \mathbb{R},$$

*where $\omega \in C^0(I \times \mathbb{R}_+)$ is nondecreasing with respect to the second variable for any fixed $t \in \mathbb{R}_+$. Let $\beta \in C^0(\mathbb{R}_+)$ be positive and satisfy*

$$\varepsilon + 2 \int_T^t h^2(s)\omega \left( s, \beta(s)h^2(s) \right) ds \le \beta(t), \quad t \in I. \tag{6.42}$$

*If y is a solution of*

$$y''' + qy' + \frac{q'}{2}y = F(t, y) \tag{6.43}$$

*satisfying the Cauchy initial conditions*

$$y(T) = C_1, \quad y'(T) = C_2, \quad y''(T) = C_3$$

*and*

$$|C_1|(1 + 4q(0)) + |C_2| + \left|\frac{C_3}{2}\right| < \varepsilon,$$

*then y is defined on $\mathbb{R}_+$ and*

$$|y(t)| \le \beta(t)h^2(t), \quad t \in I.$$

*Proof of Theorem 6.7.* Let $y$ be a solution of (6.24). Then $y(T) = C_1$, $y'(T) = C_2$, and $y''(T) = C_3$ for some constants $C_1$, $C_2$, and $C_3$, and we set

$$\varepsilon = |C_1|(1 + 4q(T)) + |C_2| + \frac{|C_3|}{2} + 1.$$

By (6.36) (also see Theorem 5.8 with $n = 1$ and $a(t) \equiv 1$), all solutions of (6.25) are bounded. Moreover, since (6.25) is of the limit–circle type, we have

$$\int_T^\infty h^4(s) \, ds < \infty. \tag{6.44}$$

Clearly, (6.24) is equivalent to (6.43) on $[T, \infty)$ where

$$F(t, x) = \left(-r(t) + \frac{q'(t)}{2}\right)x.$$

Then, with

$$\omega(t, x) = Kx \quad \text{and} \quad \beta(t) \equiv \beta_0 = \frac{2 + \delta}{\delta}\varepsilon$$

for $t \in [T, \infty)$ and $x \in \mathbb{R}_+$, we see that

$$|F(t, x)| = \left|\left(-r(t) + \frac{q'(t)}{2}\right)x\right| \le K|x| = \omega(t, |x|).$$

In addition,

$$\varepsilon + 2\int_T^t h^2(s)\,\omega\left(s, \beta(s)h^2(s)\right) ds = \varepsilon + 2K\beta_0 \int_T^t h^4(s)\,ds \le \varepsilon + \beta_0 \frac{2}{2 + \delta} = \beta_0,$$

so (6.42) holds. By Lemma 6.5, $|y(t)| \leq \beta_0 h^2(t)$, so (6.44) yields

$$\int_T^\infty y^2(s)\, ds \leq \beta_0^2 \int_T^\infty h^4(s)\, ds < \infty,$$

and this completes the proof of the theorem.                                          □

To show the applicability of Theorem 6.7, we have the following corollary.

**Corollary 6.2.** *If* (6.37) *and* (6.38) *hold and*

$$\lim_{t \to \infty} \left( r(t) - \frac{q'(t)}{2} \right) = 0, \tag{6.45}$$

*then equation* (6.24) *is of the limit–circle type.*

*Proof.* As we said above, by Theorem 1.3 with $a(t) \equiv 1$, (6.37) guarantees that equation (6.25) is of the limit–circle type. Condition (6.45) ensures the existence of $T \geq 0$ so that (6.41) holds.                                          □

**Theorem 6.8.** *Let*

$$\int_0^\infty \frac{2r(u) - q'(u)}{q(u)}\, du < \infty$$

*and suppose there are constants $\varepsilon > 0$ and $T > 0$ such that*

$$2r(t) - q'(t) \geq \varepsilon \quad and \quad q^{-\frac{1}{4}} \text{ is convex on } [T, \infty).$$

*Then equation* (6.24) *is of the limit–circle type.*

*Proof.* First, we prove that

$$\lim_{t \to \infty} q(t) = \infty. \tag{6.46}$$

Since $q^{-\frac{1}{4}}$ is convex on $[T, \infty)$, $\lim_{t \to \infty} q(t) = q_0 \geq 0$ exists. Moreover, the hypotheses of the theorem imply

$$\infty > \int_T^\infty \frac{2r(t) - q'(t)}{q(t)}\, dt \geq \int_T^\infty \frac{\varepsilon\, dt}{q(t)},$$

so $q_0 = \infty$.

Let $y$ be a proper nonoscillatory solution of (6.24). Then Theorems $3, 5$, and 7(ii) in [17] imply $y \in L^2$.

Let $y$ be a proper oscillatory solution of (6.24). Then (6.46) and Theorem 3.18 in [68] imply that every oscillatory solution $y$ of (6.24) has a bounded derivative $y'$ on $\mathbb{R}_+$. Define the function $Z(t) = -2y''y + (y')^2 - qy^2$. Then,

$$Z'(t) = 2ryf(y) - q'y^2 \geq (2\alpha_1 r - q')y^2 \geq \varepsilon y^2,$$

for $t \geq T$, and so $Z$ is nondecreasing. Let $\{t_k\} \to \infty$ be a sequence of zeros of $y$. Then,

$$Z(t_k) = (y'(t_k))^2 \leq M, k = 1, 2, \ldots,$$

so $Z(t) \leq M$ for $t \geq T$. Integrating $Z'$, we obtain

$$\int_T^\infty y^2(t)\, dt \leq \frac{1}{\varepsilon} \int_T^\infty Z'(t)\, dt = \frac{1}{\varepsilon}(Z(\infty) - Z(T)) < \infty.$$

Thus, any oscillatory solution belongs to $L^2$ and the conclusion is proved.   □

We conclude this section with an application of Theorem 6.8.

**Corollary 6.3.** *Suppose there are positive constants $\varepsilon$, $\varepsilon_1$, $q_0$, $b_0$, and $m$, and a function $B \in C^0(\mathbb{R}_+)$ such that*

$$q(t) = \frac{2q_0 t^{m+1}}{m+1}, \quad r(t) = q_0 t^m + B(t), \quad and \quad \varepsilon \leq B(t) \leq b_0 t^{m-\varepsilon_1}.$$

*Then equation (6.24) is of the limit–circle type.*

**Remark 6.4.** An excellent discussion of asymptotic and oscillatory behavior of solutions of third order linear differential equations can be found in the book of Greguš [68].

## 6.3. Nonlinear Three-Term Equations

In this section, we give limit–point/limit–circle criteria for the equation

$$y''' + q(t)y' + r(t)f(y) = 0, \tag{6.47}$$

where $q \in C^0(\mathbb{R}_+)$, $r \in L_{loc}(\mathbb{R}_+)$, and $f : \mathbb{R} \to \mathbb{R}$ is continuous with $xf(x) \geq 0$ for all $x$.

If $h$ is a positive solution of (6.25), then

$$y''' + q(t)y' = \frac{1}{h}\left(h^2\left(\frac{1}{h}y'\right)'\right)'.$$

Hence, (6.47) can be rewritten as

$$\left(h^2(t)\left(\frac{1}{h(t)}y'\right)'\right)' = -h(t)r(t)f(y), \tag{6.48}$$

which is in the form of (6.1) (see [25, Section 3]). Also, if $r(t) < 0$, then (6.2) holds with $\alpha = 0$.

**Corollary 6.4.** *Suppose* $xf(x) > 0$ *for all* $x \neq 0$, $q(t) \leq -L \leq 0$, *and* $\mu(t)r(t)$
$\leq K < 0$, *where* $\mu(t) = e^{\sqrt{L}t}$ *for* $L > 0$ *and* $\mu(t) = t$ *for* $L = 0$. *Suppose there
exist* $M > 0$ *and* $\beta$ *such that*

$$|f(x)| \leq \frac{1}{|x|^\beta} \quad for \ |x| \geq M,$$

*where* $\beta > 1$ *for* $L > 0$ *and* $\beta > \frac{3}{2}$ *for* $L = 0$. *Then equation* (6.47) *is of the
nonlinear limit–circle type.*

*Proof.* We apply Theorem 6.2 to (6.48). To ensure that (6.3) holds, we choose
$h$ to be a nondecreasing solution of equation (6.25), which we can do since this
equation is nonoscillatory. Let $h$ satisfy the initial conditions

$$h(0) \geq 1, \quad h'(0) \geq \sqrt{L}.$$

Next, we verify that (6.11) holds for such an $h$. Since $q(t) \leq -L$, we can compare
equation (6.25) to the equation $v'' - Lv = 0$ and obtain

$$h(t) \geq \begin{cases} e^{\sqrt{L}t}, & \text{if } L > 0, \\ t, & \text{if } L = 0. \end{cases}$$

With $a_1(t) = h(t)$, $a_2(t) = 1/h^2(t)$, and taking the above estimate into consider-
ation, we have

$$\int_1^\infty \frac{du}{\left(\int_0^u h(w) \int_0^w \frac{1}{h^2(\tau)} \, d\tau dw\right)^{\beta-1}} < \infty,$$

i.e., (6.11) is satisfied. Clearly, the assumption $-h(t)r(t) = -\mu(t)r(t) \geq |K| > 0$
is satisfied. $\qquad\square$

An application of Theorem 6.1 yields the following limit–point result.

**Corollary 6.5.** *Let equation* (6.25) *be nonoscillatory,* $r(t) < 0$, *and let there exist
$M > 0$ and $C > 0$ such that*

$$\frac{1}{|x|} \leq |f(x)| \leq C|x| \quad for \ |x| \geq M.$$

*Then equation* (6.47) *is of the nonlinear limit–point type.*

Throughout the remainder of this section, we restrict our attention to proper solutions of (6.47), i.e., to those solutions of (6.47) which are defined on $\mathbb{R}_+$ and are nontrivial in any neighborhood of infinity.

It is convenient to let $\mathcal{O}$ and $\mathcal{N}$ denote the sets of oscillatory and nonoscillatory solutions of (6.47), respectively, and define the following classes of solutions (see, for example, [7]):

$$W_1 = \{y \in \mathcal{N} : y' \text{ is oscillatory}\},$$

$$W_2 = \{y \in \mathcal{N} : y(t)y'(t) < 0 \text{ for large } t \text{ and } y'' \text{ is oscillatory}\},$$

$$\mathcal{N}_0 = \{y \in \mathcal{N} : y(t)y'(t) < 0 \text{ and } y(t)y''(t) > 0 \text{ for large } t\},$$

$$\mathcal{N}_1 = \{y \in \mathcal{N} : y(t)y'(t) > 0 \text{ and } y(t)y''(t) > 0 \text{ for large } t\}.$$

Finally, we recall that the solutions in the class $\mathcal{N}_0$ are called Kneser solutions, and those in the classes $W_1$ or $W_2$ are called weakly oscillatory solutions.

In what follows, we need to impose a growth condition on $f$, namely, that there exists constants $0 < \alpha_1 \le \alpha$ such that

$$\alpha_1|u| \le |f(u)| \le \alpha|u| \text{ for } u \in \mathbb{R}. \tag{6.49}$$

Observe that this implies that the nonlinear limit–circle property is equivalent to the square integrability of solutions.

In view of condition (6.49), every solution (6.47) is defined on $\mathbb{R}_+$ (see [2, Theorem 1.4.]), and moreover, a nontrivial solution is nontrivial in any neighborhood of infinity (see [1; Theorem 1.3]). The structure of the solutions of (6.47) is described in the following lemma [7, Theorem 1].

**Lemma 6.6.** *Let $y$ be a nontrivial solution of (6.47). Then $y \in \mathcal{O} \cup W_1 \cup W_2 \cup \mathcal{N}_0 \cup \mathcal{N}_1$.*

**Theorem 6.9.** *Suppose (6.49) holds and let $T > 0$, $M_1 > 0$, and $M > 0$ be such that one of the following conditions holds:*

(i) $q(t) \le M_1 t$ *and* $q'(t) \ge 2\alpha r(t)$ *for* $t \ge T$;

(ii) $q \in C^3(\mathbb{R}_+)$, $q(t) > 0$, $M \ge 0$,

$$1 + \left(\frac{t}{q(t)}\right)''' - 2\alpha \frac{tr(t)}{q(t)} \ge 0, \qquad 0 \le -\left(\frac{t}{q(t)}\right)' \le M,$$

$$\text{and} \quad \left(\frac{t}{q(t)}\right)'' \le M_1 t \qquad \text{for } t \ge T. \tag{6.50}$$

*Then equation (6.47) is of the nonlinear limit–point type, i.e., there is a solution of (6.47) that does not belong to $L^2(\mathbb{R}_+)$.*

*Proof.* With $R(t) \equiv 1$ and $M = 0$ in case (i), and $R(t) \equiv \frac{1}{q(t)}$ in case (ii), we define

$$Z(t) = Ry'y - \frac{3}{2} \int_T^t R(s)(y'(s))^2 \, ds - R'y^2 + \int_T^t [R(s)q(s) + 3R''(s)] \frac{y^2(s)}{2} \, ds.$$

Then,

$$Z'(t) = Ry''y - R \frac{(y')^2}{2} - R'y'y + [Rq + R''] \frac{y^2}{2}$$

and

$$Z''(t) = -rRf(y)y + [(qR)' + R'''] \frac{y^2}{2} - \frac{3}{2} R'(y')^2$$

$$\geq [-2\alpha r R + (qR)' + R'''] \frac{y^2}{2} - \frac{3}{2} R'(y')^2. \qquad (6.51)$$

Moreover, (6.49) and the hypotheses of the theorem show that in both cases we have $Z''(t) \geq 0$ for $t \geq T$. Let $y$ be a solution of (6.47) such that $Z'(T) > 0$. Clearly, $y$ is nontrivial in any neighborhood of $\infty$, for otherwise $Z \equiv 0$ in this neighborhood and that contradicts $Z'(T) > 0$ and $Z'$ nondecreasing. Thus, $y$ is a proper solution. It is easy to see that $y$ must be nonoscillatory since if there is a sequence $\{t_k\} \to \infty$ of zeros of $y$, then $Z'(t_k) \leq 0, k = 1, 2, \ldots$. In addition, since $Z'$ is nondecreasing, we would have $\lim_{t \to \infty} Z'(t) \leq 0$, which contradicts $Z'(T) > 0$ and $Z''(t) \geq 0$. Hence, by Lemma 6.6,

$$y \in W_1 \cup W_2 \cup \mathcal{N}_0 \cup \mathcal{N}_1.$$

First note that if $y \in \mathcal{N}_1$, then $\int_T^\infty y^2(t) \, dt = \infty$, and so the conclusion of the theorem holds. For the remainder of the possibilities, we proceed by contradiction. Suppose $y$ satisfies

$$\int_0^\infty y^2(t) \, dt < \infty$$

and say $y(t) > 0$ for $t \geq T_1 \geq T$. Then, there exists $\tau \geq T_1$ such that

$$\int_\tau^\infty y^2(s) \, ds \leq \frac{Z'(T)}{1 + 3M_1}. \qquad (6.52)$$

Since $Z'$ is nondecreasing for $t \geq T$, (6.50)–(6.52) yield

$$Z'(T)(t - T) \leq Z(t) - Z(T)$$

$$\leq R(t)y'(t)y(t) + My^2(t) + \frac{1 + 3M_1}{2} t \int_\tau^\infty y^2(s)\,ds$$

$$+ \int_T^\tau \left[R(s)q(s) + 3R''(s)\right] \frac{y^2(s)}{2}\,ds - Z(T)$$

$$\leq R(t)y'(t)y(t) + My^2(t) + \frac{Z'(T)t}{2} + K, \quad (6.53)$$

for $t \geq \tau$ and some constant $K$. If $y \in \mathcal{N}_0 \cup W_2$, then $y$ is decreasing and $yy' < 0$ for large $t$, and so we have a contradiction.

Finally, suppose $y \in W_1$. Now (6.52) implies $\liminf_{t \to \infty} y(t) = 0$ and there exists a sequence $\{t_k\} \to \infty$ of zeros of $y'$ such that $\lim_{k \to \infty} y(t_k) = 0$ (for example, a sequence of local minimums of $y$). We again obtain a contradiction by taking $t = t_k$ in (6.53) with $k$ large.                                                               □

The following theorem gives us a limit–point type result for the case where the functions $q$ and $r$ are small.

**Theorem 6.10.** *Let* (6.49) *hold and suppose there exist constants* $T \geq 0$, $M \geq 0$, *and* $M_1 \geq 0$ *such that for* $t \geq T$,

$$q(t) \leq M, r(t) \leq M_1 t^{\frac{1}{2}}, \text{ and } 2\alpha_1 r(t) \geq q'(t).$$

*Then equation* (6.47) *is of the nonlinear limit–point type.*

*Proof.* For a solution $y$ of (6.47), we define

$$Z(t) = -2y'(t)y(t) + 3\int_T^t \left(y'(s)\right)^2\,ds - \int_T^t q(s)y^2(s)\,ds. \quad (6.54)$$

Then,

$$Z'(t) = -2y''y + (y')^2 - qy^2 \quad (6.55)$$

and

$$Z''(t) = 2ryf(y) - q'y^2 \geq (2\alpha_1 r - q')y^2 \geq 0$$

for $t \geq T$. Thus, $Z'$ is nondecreasing.

Let $y$ be a solution of (6.47) such that $Z'(T) > 0$. Once again, $y$ is proper, and Lemma 6.6 implies $y \in \mathcal{O} \cup W_1 \cup W_2 \cup \mathcal{N}_0 \cup \mathcal{N}_1$. Integrating, we obtain

$$\frac{Z'(T)}{2} t \le Z'(T)(t - T) \le Z(t) - Z(T)$$

$$\le -2y'(t)y(t) + 3 \int_T^t (y'(s))^2 \, ds \qquad (6.56)$$

for $t \ge T_1$ for some $T_1 \ge T$. Let $y \in W_2 \cup \mathcal{N}_0$; then $y(t)y'(t) < 0$ for $t \ge T_2 \ge T_1$, and there exists a sequence $\{t_k\} \to \infty$ such that $\lim_{k \to \infty} y'(t_k) = 0$ and $y''(t_k)y(t_k) \ge 0$. (If $y \in \mathcal{N}_0$, then any sequence tending to $\infty$ will work, and if $y \in W_2$, then it is a sequence of local maxima of $y'(t) \operatorname{sgn} y(t)$.) Hence, $\lim_{k \to \infty} Z'(t_k) \le 0$, and since $Z'$ is nondecreasing, $Z'(t) \le 0$ for large $t$. This contradicts $Z'(T) > 0$ and $Z'' \ge 0$.

If $y \in \mathcal{N}_1$, the conclusion of the theorem clearly holds. Now let $y \in \mathcal{O} \cup W_1$ and let $\{t_k\} \to \infty$, be a sequence of zeros of $y'$. Then, (6.56) yields

$$\frac{Z'(T)}{2} t_k \le 3 \int_T^{t_k} (y'(s))^2 \, ds, \quad k = 1, 2, \dots \qquad (6.57)$$

for $t_k \ge T_2$. Let $J_k = [T, t_k], k = 1, 2, \dots, v \in C^0[T, \infty)$, and $\|v\|_k = \left( \int_T^{t_k} v^2(s) \, ds \right)^{\frac{1}{2}}$. Suppose, to the contrary, that $\int_T^\infty y^2(s) \, ds < \infty$. Then, (6.49) yields the existence of constants $K$ and $K_1$ such that

$$\int_T^\infty y^2(s) \, ds = K^2 < \infty \text{ and } \int_T^\infty [f(y(s))]^2 \, ds = K_1^2 < \infty. \qquad (6.58)$$

Set $\varepsilon = \frac{\varepsilon_1}{2M_1 K_1 + M\varepsilon_1}$, where $\varepsilon_1 = \sqrt{\frac{Z'(T)}{6}}$. By [90, Theorem 1.2], there exists a constant $K_2$ such that

$$\|y'\|_k \le \varepsilon \|y'''\|_k + K_2 \|y\|_k, \quad k = 1, 2, \dots. \qquad (6.59)$$

From (6.47) and Minkowski's inequality, we obtain

$$\|y'''\|_k = \|-qy' - rf(y)\|_k \le \|qy'\|_k + \|rf(y)\|_k. \qquad (6.60)$$

From (6.58)–(6.60), we have

$$\|y'\|_k \le \varepsilon \left( M \|y'\|_k + M_1 t_k^{\frac{1}{2}} \|f(y)\|_k \right) + K_2 \|y\|_k,$$

so, using (6.57),

$$2\varepsilon M_1 K_1 t_k^{\frac{1}{2}} \le (1 - \varepsilon M) \|y'\|_k \le \varepsilon M_1 K_1 t_k^{\frac{1}{2}} + K_2 K.$$

This contradiction for large $k$ completes the proof of the theorem.                $\square$

**Example 6.1.** Consider the equation

$$y''' + qt^m y' + rt^n f(y) = 0 \tag{6.61}$$

where $q > 0$ and $r > 0$ are constants and (6.49) holds. By parts (i) (for $0 < m \leq 1$) and (ii) (for $m > 1$) of Theorem 6.9, we see that equation (6.61) is of the nonlinear limit–point type if any of the following conditions hold:

  (a) $m > n + 1$;

  (b) $m = n + 1, 0 < m \leq 1$, and $mq > 2\alpha r$;

  (c) $m = n + 1, m > 1$, and $q > 2\alpha r$;

  (d) $m \leq 0$ and $n \leq 1/2$.

If we apply Corollary 6.2 to equation (6.61), we obtain that (6.61) is of the limit–circle type provided $m > 2, n = m - 1, mq = 2r$, and $f(x) \equiv x$. Observe that under these restrictions on $n$ and $m$, (6.61) has the form of Appel's equation.

**Open Problems.**

**Problem 6.1.** *What conditions on the coefficients in the third order equations (6.1), (6.24), or (6.47) must be added so that a limit–circle type equation has an oscillatory solution?*

**Problem 6.2.** *We know that, in the second order case, the limit–circle property is not enough to guarantee that all solutions are bounded. What is the situation for third order equations?*

**Problem 6.3.** *What conditions on the coefficients in the third order equation of the limit–circle type must be added so that all proper solutions tend to zero as t tends to infinity?*

**Notes.** Theorems 6.1–6.4 are taken from Bartušek and Došlá [11]. Theorems 6.5 appears in Bartušek and Došlá [11]. Theorem 6.6 is due to Došlá [36], Theorems 6.7–6.10 and Example 6.1 are due to Bartušek and Graef [18].

# Chapter 7

# Fourth Order Differential Equations

In addition to discussing general fourth order nonlinear equations with quasiderivatives, we consider sublinear equations in self-adjoint form, a two term nonlinear equation, and then fourth order linear equations.

## 7.1. Equations with Quasiderivatives

We consider the equation

$$y^{[4]} \equiv \left( \frac{1}{a_3(t)} \left( \frac{1}{a_2(t)} \left( \frac{1}{a_1(t)} y' \right)' \right)' \right)' = r(t) f(y, y^{[1]}, y^{[2]}, y^{[3]}), \qquad (7.1)$$

where

$$y^{[i]} = \frac{1}{a_i(t)} (y^{[i-1]})', \quad i = 1, 2, 3, \quad y^{[0]} = y.$$

We assume that the functions $a_i \in C^0(\mathbb{R}_+)$ are positive $(i = 1, 2, 3)$, $r \in L_{\text{loc}}(\mathbb{R}_+)$, $f : \mathbb{R}^4 \to \mathbb{R}$ is continuous, $x_1 f(x_1, x_2, x_3, x_4) \geq 0$ on $\mathbb{R}^4$, and either

$$r(t) \leq 0 \quad \text{on } \mathbb{R}_+ \qquad (7.2)$$

or

$$r(t) \geq 0 \quad \text{on } \mathbb{R}_+. \qquad (7.3)$$

We start with the case $r(t) \geq 0$.

**Theorem 7.1.** *Let $r(t) \geq 0$ and suppose there exist constants $K > 0$ and $C > 0$ such that*

$$\frac{1}{|x_1|} \leq |f(x_1, x_2, x_3, x_4)| \leq C|x_1| \quad \text{for } |x_1| \geq K. \tag{7.4}$$

*Then (7.1) is of the nonlinear limit-point type.*

*Proof.* Let $y$ be a solution of equation (7.1) with the initial conditions

$$y(0) \geq K, \quad y^{[i]}(0) > 0, \quad i = 1, 2, 3.$$

According to (7.1), (7.3), and the definition of quasiderivatives, we have $y^{[j]}(t) \geq 0$ for $j \in \{0, 1, 2, 3, 4\}$ on $J$, so, $y^{[j-1]}$ is nondecreasing for $j \in \{1, 2, 3, 4\}$. It is clear that $y$ is nonoscillatory and that it is either proper or singular. By Theorem 2.1, we have that (7.4) ensures that $y$ cannot be a singular solution, so $y$ is a proper solution, and $y^{[i]} > 0, i = 0, 1, 2, 3$, are nondecreasing on $\mathbb{R}_+$. Hence,

$$\int_0^\infty y(t) f(y(t), y^{[1]}(t), y^{[2]}(t), y^{[3]}(t)) \, dt \geq \int_K^\infty dt = \infty,$$

and so $y$ is of the nonlinear limit–point type.                                    □

In what follows, we treat the case $r(t) \leq 0$. Define the functions $F$ and $w$ by

$$F(t) = -yy^{[3]} + \frac{a_1}{a_3} y^{[1]} y^{[2]} - \frac{1}{2a_2} \left(\frac{a_1}{a_3}\right)' (y^{[1]})^2$$

and

$$w(t) = -yy^{[2]} + \frac{a_1}{a_2} (y^{[1]})^2 - \int_0^t \left[\left(\frac{a_1}{a_2}\right)' + \frac{a_3}{2a_2} \left(\frac{a_1}{a_3}\right)'\right] (y^{[1]})^2 \, ds,$$

where $y$ is a solution of (7.1). Some useful properties of these functions are given in the following lemma.

**Lemma 7.1.** *Let (7.2) hold and*

$$\left[\frac{1}{a_2} \left(\frac{a_1}{a_3}\right)'\right]' \leq 0. \tag{7.5}$$

*Then,*

$$F'(t) = -yy^{[4]} - \frac{1}{2}\left[\frac{1}{a_2}\left(\frac{a_1}{a_3}\right)'\right]' (y^{[1]})^2 + \frac{a_1 a_2}{a_3}(y^{[2]})^2 \geq 0$$

*and*

$$w' = a_3(t) F(t).$$

*Proof.* Using (7.2) and (7.5), we have

$$F' = -a_1 y^{[1]} y^{[3]} - yy^{[4]} + \left(\frac{a_1}{a_3}\right)' y^{[1]} y^{[2]} + \frac{a_1 a_2}{a_3} (y^{[2]})^2$$

$$+ a_1 y^{[1]} y^{[3]} - \left[\frac{1}{2a_2}\left(\frac{a_1}{a_3}\right)'\right]' (y^{[1]})^2 - \left(\frac{a_1}{a_3}\right)' y^{[1]} y^{[2]} \geq 0.$$

Moreover,

$$\frac{w'}{a_3} = \frac{1}{a_3}\left[-a_3 yy^{[3]} - a_1 y^{[1]} y^{[2]} + \left(\frac{a_1}{a_2}\right)' (y^{[1]})^2 + 2a_1 y^{[1]} y^{[2]}\right.$$

$$\left. - \left(\frac{a_1}{a_2}\right)' (y^{[1]})^2 - \frac{a_3}{2a_2}\left(\frac{a_1}{a_3}\right)' (y^{[1]})^2\right] = F.$$

$\square$

**Lemma 7.2.** *Suppose (7.5) holds and there exists $t_0 > 0$ such that*

$$\left(\frac{a_1}{a_2}\right)' + \frac{a_3}{2a_2}\left(\frac{a_1}{a_3}\right)' \geq 0 \ \text{and} \ \frac{a_2}{a_1}\int_0^t a_3(s)ds \geq c > 0 \ \text{for } t \geq t_0. \tag{7.6}$$

*Let $M > 0$ be an arbitrary constant and let $y$ be an oscillatory solution of (7.1) having initial conditions satisfying*

$$F(0) \geq \frac{M^2}{c} \ \text{and} \ w(0) > 0. \tag{7.7}$$

*Then, for any relative extrema $\tau \geq t_0$ of $y^{[1]}$, i.e., $y^{[2]}(\tau) = 0$, we have*

$$|y^{[1]}(\tau)| \geq M.$$

*Proof.* From the definition of $w$ and (7.7), we have

$$w \leq -yy^{[2]} + \frac{a_1}{a_2}(y^{[1]})^2.$$

From Lemma 7.1 and (7.1), we have

$$\frac{M^2}{c}\int_0^t a_3(s)ds \leq \int_0^t a_3(s)F(s)\,ds \leq w(t) \leq -yy^{[2]} + \frac{a_1}{a_2}(y^{[1]})^2,$$

and it follows that

$$(y^{[1]}(\tau))^2 \geq \frac{a_2}{a_1}yy^{[2]}|_\tau + \frac{M^2}{c}\frac{a_2(\tau)}{a_1(\tau)}\int_0^\tau a_3(s)\,ds \geq M^2 \ \text{for } \tau \geq t_0.$$

$\square$

To prove that (7.1) is of the nonlinear limit–point type, it suffices to show that it has a solution $y$ satisfying

$$\int_0^\infty y(t) f(y(t), y^{[1]}(t), y^{[2]}(t), y^{[3]}(t)) dt = \infty.$$

As we will show, such solutions will be one of the following two types.

**Type I.** $y$ is oscillatory and there exist increasing sequences $\{t_k^i\}$, $i = 0, 1, 2, 3$, $\{\bar{t}_k^3\}$, $k \in \mathbb{N}$, tending to $\infty$ such that

$$\tau \le t_k^1 < t_k^0 < t_k^3 \le \bar{t}_k^3 < t_k^2 < t_{k+1}^1, k \in \mathbb{N}, \ t_1^1 = \tau,$$

$$y^{[i]}(t_k^i) = 0, i = 0, 1, 2, \ \ y^{[3]}(t) \equiv 0 \text{ for } t \in [t_k^3, \bar{t}_k^3],$$

$$yy^{[1]} \begin{cases} > 0 & \text{on } (t_k^0, t_{k+1}^1), \\ < 0 & \text{on } (t_k^1, t_k^0), \end{cases} \qquad yy^{[2]} \begin{cases} > 0 & \text{on } (t_k^1, t_k^2), \\ < 0 & \text{on } (t_k^2, t_{k+1}^1), \end{cases}$$

$$y^{[1]}y^{[3]} \begin{cases} > 0 & \text{on } (t_k^1, t_k^3), \\ < 0 & \text{on } (\bar{t}_k^3, t_{k+1}^1). \end{cases}$$

**Type II.** $y$ is nonoscillatory and there exists $\tau \in \mathbb{R}_+$ such that

$$yy^{[j]} > 0 \ \ \text{on } [\tau, \infty), \ \ j = 1, 2, \ \ yy^{[3]} \ge 0 \ \ \text{on } [\tau, \infty),$$

and $|y^{[3]}|$ is nonincreasing on $[\tau, \infty)$.

**Lemma 7.3.** ([4, Lemmas 2 and 9]) *Let $a_i \in C^1(\mathbb{R}_+)$, $i = 1, 2, 3$, $\frac{a_3}{a_1} \in C^2(\mathbb{R}_+)$, and let $y$ be a solution of equation (7.1) defined on $\mathbb{R}_+$ with the initial conditions*

$$y(\tau) > 0, \ \ y^{[1]}(\tau) > 0, \ \ y^{[2]}(\tau) = 0, \ \ y^{[3]}(\tau) < 0, \tag{7.8}$$

*where $\tau \in \mathbb{R}_+$. Then $y$ is either of Type I or Type II.*

The following lemma will be needed to prove our next nonlinear limit–point result.

**Lemma 7.4.** *Suppose that either*

$$a_2'(t) \ge 0 \ \ and \ \ a_1'(t) \le 0 \tag{7.9}$$

*or*

$$a_2'(t) \ge 0, \ a_3'(t) \ge 0, \ \ and \ for \ some \ T \ge 0, \ \ \frac{\max_{[T,\infty)} a_1(t)}{\min_{[T,\infty)} a_1(t)} \le 2. \tag{7.10}$$

Let $M > 0$ and let $y$ be an oscillatory solution of equation (7.1) with initial conditions

$$y(t_0) > 0, \quad y^{[1]}(t_0) \geq M, \quad y^{[2]}(t_0) = 0, \quad y^{[3]}(t_0) < 0, \tag{7.11}$$

where $t_0 = 0$ in case (7.9) holds, and $t_0 \geq T$ in case (7.10) holds. Then, at any relative extrema $\tau \geq t_0$ of $y^{[1]}$, we have

$$|y^{[1]}(\tau)| \geq M.$$

*Proof.* Let $y$ be an oscillatory solution of (7.1) satisfying (7.11). By Lemma 7.3, $y$ is a Type I solution of (7.1), so let $\{\bar{\tau}_k\}_{k=1}^{\infty}$ be the increasing sequence of relative extrema of $y^{[1]}$, i.e., $y^{[2]}(\bar{\tau}_k) = 0$, with $\bar{\tau}_1 = t_0$. We will prove that $\{|y^{[1]}(\bar{\tau}_k)|\}$ is increasing, and so the conclusion of the lemma follows.

To the contrary, suppose that for some successive $\tau_1 < \tau_2$ from $\{\bar{\tau}_k\}_{k=1}^{\infty}$ we have

$$|y^{[1]}(\tau_1)| \geq |y^{[1]}(\tau_2)|. \tag{7.12}$$

For simplicity, we denote by $t_1, t_2, t_3, t_4$ successively, the zeros of $y^{[2]}$, $y^{[1]}$, $y$, and $y^{[2]}$, i.e.,

$$y^{[2]}(t_1) = 0, \quad y^{[1]}(t_2) = 0, \quad y(t_3) = 0, \quad y^{[2]}(t_4) = 0,$$

where $\tau_1 = t_1 < t_2 < t_3 < t_4 = \tau_2$. Then, since $y$ is a Type I solution:

($\alpha$) $y^{[2]} \neq 0$ on $(t_1, t_4)$;

($\beta$) $y^{[1]}$ sgn $y > 0$ on $(t_1, t_2)$, $y^{[1]}$ sgn $y < 0$ on $(t_2, t_3)$, and $y^{[1]}$ sgn $y$ is decreasing on $(t_1, t_3)$;

($\gamma$) $y^{[2]}$ sgn $y < 0$ and decreasing on $(t_1, t_3)$;

($\delta$) $y^{[3]}$ sgn $y < 0$ and nonincreasing on $(t_1, t_3)$.

Now (7.12), Lemma 7.3, and ($\alpha$) imply

$$|y^{[1]}(t_1)| \geq |y^{[1]}(t_4)| > |y^{[1]}(t_3)| > 0,$$

so

$$1 < \frac{|y^{[1]}(t_1)|}{|y^{[1]}(t_3)|}. \tag{7.13}$$

Denote by $\Delta_{ij}$ the length of the interval $[t_i, t_j]$, $i, j \in \{1, 2, 3, 4\}$.

Since $a_2' \geq 0$, $(y^{[1]})' = a_2 y^{[2]}$, and $(y^{[1]})'' = a_2 a_3 y^{[3]} + a_2' y^{[2]}$, we have $(y^{[1]})''$ sgn $y \leq 0$ on the interval $[t_1, t_3]$. This implies that $y^{[1]}$ sgn $y$ is concave down on $[t_1, t_3]$, and moreover, using ($\beta$),

$$\int_{t_2}^{t_3} |y^{[1]}| \leq \frac{1}{2} |y^{[1]}(t_3)| \Delta_{23}, \quad \int_{t_1}^{t_2} |y^{[1]}| \geq \frac{1}{2} |y^{[1]}(t_1)| \Delta_{12} > 0.$$

It follows that

$$|y(t_2)| = \int_{t_2}^{t_3} |y'| = \int_{t_2}^{t_3} |a_1 y^{[1]}|$$

$$\leq \max_{[t_2,t_3]} a_1(t) \int_{t_2}^{t_3} |y^{[1]}| \leq \frac{1}{2} \max_{[t_2,t_3]} a_1(t) |y^{[1]}(t_3)| \Delta_{23}, \quad (7.14)$$

and

$$|y(t_2)| > |y(t_2)| - |y(t_1)| = \int_{t_1}^{t_2} a_1 |y^{[1]}|$$

$$\geq \min_{[t_1,t_2]} a_1(t) \int_{t_1}^{t_2} |y^{[1]}| \geq \frac{1}{2} \min_{[t_1,t_2]} a_1(t) |y^{[1]}(t_1)| \Delta_{12}. \quad (7.15)$$

In addition, we have

$$|y^{[1]}(t_3)| = \int_{t_2}^{t_3} a_2 |y^{[2]}| \geq a_2(t_2) |y^{[2]}(t_2)| \Delta_{23}, \quad (7.16)$$

$$|y^{[1]}(t_1)| = \int_{t_1}^{t_2} a_2 |y^{[2]}| \leq a_2(t_2) \int_{t_1}^{t_2} |y^{[2]}| \leq a_2(t_2) |y^{[2]}(t_2)| \Delta_{12}. \quad (7.17)$$

(i) Suppose (7.9) holds. Combining (7.14) and (7.15), respectively (7.16) and (7.17), with (7.13), we obtain

$$1 < \frac{\Delta_{12}}{\Delta_{23}} < 1,$$

which is a contradiction.

(ii) Suppose (7.10) holds. We proceed in a similar way using the fact that $(y^{[2]})' = a_3 y^{[3]}$ and $(y^{[2]})'' = a_3 a_4 y^{[4]} + a_3' y^{[3]}$. Since $a_3' \geq 0$, $y^{[2]} \operatorname{sgn} y$ is concave down on $[t_1, t_2]$, so

$$\int_{t_1}^{t_2} |y^{[2]}| \leq \frac{1}{2} |y^{[2]}(t_2)| \Delta_{12},$$

and hence

$$|y^{[1]}(t_1)| \leq \frac{1}{2} a_2(t_2) |y^{[2]}(t_2)| \Delta_{12}.$$

This, together with (7.13)–(7.16), yields

$$1 < \frac{|y^{[1]}(t_1)|}{|y^{[1]}(t_3)|} \leq \frac{\Delta_{12}}{2\Delta_{23}} < \frac{\max_{[t_2,t_3]} a_1}{2 \min_{[t_1,t_2]} a_1} \leq 1,$$

which is a contradiction. This completes the proof of the lemma.     □

**Remark 7.1.** Assumption (7.10) can be applied, for example, when $a_1(t) = c + \sin t$, where $c > 3$; clearly, (7.9) does not hold in this case.

Our next lemma gives a Kolmogorov–Horny type inequality similar to what we had for third order equations in Lemma 6.2 above.

**Lemma 7.5.** *Let $-\infty < t_1 < t_2 < \infty$ and let $a_i > 0$, $i = 1, 2, 3, 4$, be continuous functions such that the quasiderivatives $z^{[i]}$ defined by*

$$z^{[i]} = \frac{1}{a_i(t)}(z^{[i-1]})', \quad i = 1, 2, 3, 4, \quad z^{[0]} = z$$

*are continuous for $i = 0, 1, 2, 3$ and $z^{[4]} \in L_{loc}[t_1, t_2]$. Suppose that $z^{[i]}$, $i = 1, 2, 3$, each has a zero in $[t_1, t_2]$, let*

$$c_i = \max_{t_1 \leq t \leq t_2} \frac{a_{i+1}(t)}{a_i(t)}, \quad i = 1, 2, 3,$$

*and let*

$$v_i = \max_{t_1 \leq t \leq t_2} |z^{[i]}(t)|, \quad i = 0, 1, 2, 3 \text{ and } v_4 \geq |z^{[4]}| \text{ a. e. on } [t_1, t_2].$$

*Then,*

$$v_1 \leq K v_0^{\frac{3}{4}} v_4^{\frac{1}{4}},$$

*where $K = 2^{\frac{3}{2}} c_1^{\frac{3}{4}} c_2^{\frac{1}{2}} c_3^{\frac{1}{4}}$.*

*Proof.* By the same method of proof used to obtain a Kolmogorov–Horny type inequality for three quasiderivatives in Lemma 6.2, we have

$$v_1^2 \leq 2c_1 v_0 v_2, \quad v_2^2 \leq 2c_2 v_1 v_3, \quad v_3^2 \leq 2c_3 v_2 v_4.$$

Thus,

$$v_1^2 v_3^2 \leq 4c_1 c_3 v_0 v_2^2 v_4 \leq 8c_1 c_2 c_3 v_0 v_1 v_3 v_4, \quad \text{i.e.,} \quad v_1 v_3 \leq 8c_1 c_2 c_3 v_0 v_4.$$

In addition,

$$v_1^4 \leq 4c_1^2 v_0^2 v_2^2 \leq 8c_1^2 c_2 v_0^2 v_1 v_3 \leq 64c_1^3 c_2^2 c_3 v_0^3 v_4,$$

i.e.,

$$v_1 \leq 2^{\frac{3}{2}} c_1^{\frac{3}{4}} c_2^{\frac{1}{2}} c_3^{\frac{1}{4}} v_0^{\frac{3}{4}} v_4^{\frac{1}{4}}.$$

$\square$

We are now ready to prove our nonlinear limit–point result for the case of equation (7.1) with $r(t) \le 0$.

**Theorem 7.2.** *Suppose* (7.2) *and* (7.4) *hold,* $a_i \in C^1(\mathbb{R}_+)$, $i = 1, 2, 3$, $\frac{a_3}{a_1} \in C^2(\mathbb{R}_+)$, *there exist constants* $L_i > 0$, $i = 1, 2, 3$, *such that*

$$\frac{|r(t)|}{a_3(t)} \le L_1, \quad \frac{a_2(t)}{a_1(t)} \le L_2, \quad \frac{a_3(t)}{a_2(t)} \le L_3, \tag{7.18}$$

*and*

$$\int_0^\infty a_1(t)dt = \infty \ \text{ and } \ a_1 \text{ is bounded from above.} \tag{7.19}$$

*If either* (7.5)–(7.6), *or* (7.9), *or* (7.10) *holds, then equation* (7.1) *is of the nonlinear limit–point type.*

*Proof.* If (7.5)–(7.6) hold, let $\sigma = 0$, and if either (7.9) or (7.10) holds, let $\sigma = t_0$, where $t_0$ is as in Lemma 7.4. Let $y$ be a solution of equation (7.1) with the initial conditions

$$y(\sigma) > 0, \ y^{[1]}(\sigma) \ge M, \ y^{[2]}(\sigma) = 0, \ y^{[3]}(\sigma) < 0, \tag{7.20}$$

where $M = (2K+1)2^{3/2}L_1^{1/4}L_2^{3/4}L_3^{1/2}$. In addition, if (7.5)–(7.6) hold, we assume that

$$F(0) \ge \frac{M^2}{c}. \tag{7.21}$$

According to Theorem 2.1, assumption (7.4) ensures that this solution is defined on $\mathbb{R}_+$. According to Lemma 7.3, $y$ is proper and is either of Type I or Type II.

Let $y$ be of Type I. Without loss of generality, we consider the equation

$$\frac{1}{a_3}\left(\frac{1}{a_3}\left(\frac{1}{a_2}\left(\frac{1}{a_1}y'\right)'\right)'\right)' = \frac{c(t)}{a_3(t)}f(y, y^{[1]}, y^{[2]}, y^{[3]}),$$

and see that $y$ is also a solution of this equation. Let $\mathbb{N}_1 = \mathbb{N} \setminus \{1\} = \{2, 3, \dots\}$,

$$v_{ik} = \max_{t_{k-1}^0 \le t \le t_k^1} |y^{[i]}(t)|, \ i = 0, 1, 2, 3, \ \text{ and } \ v_{4k} = L_1(1 + v_{0k}), \ k \in \mathbb{N}_1.$$

Under the assumptions (7.18) and (7.4), Lemma 7.5 yields

$$v_{1k} \le M_0 v_{0k}^{3/4} v_{4k}^{1/4} \le M_0 L_1^{1/4} v_{0k}^{3/4}(1 + v_{0k})^{1/4} \le M_1(1 + v_{0k}), \ k \in \mathbb{N}_1, \tag{7.22}$$

where $M_0 = 2^{3/2} L_2^{1/4} L_3^{1/2}$ and $M_1 = M_0 L_1^{1/4}$. From this, we have

$$v_{0k} \ge \frac{v_{1k}}{M_1} - 1. \tag{7.23}$$

We will prove that there exists $k_0 \in \mathbb{N}_1$ such that

$$v_{0k} \geq 2K, \quad k \in \{k_0, k_0 + 1, \dots\}, \tag{7.24}$$

where $K$ is given in (7.4).

(i) Suppose (7.5) and (7.6) hold. Then, in view of (7.21), we can apply Lemma 7.2 to obtain

$$v_{1k} \geq (2K + 1)M_1 = M, \quad k \in \{k_0, k_0 + 1, \dots\},$$

where $k_0$ is such that $t_{k_0}^1 \geq t_0$ ($t_0$ is from (7.6)). Now (7.24) follows from (7.23).

(ii) Suppose (7.9) or (7.10) holds. Applying Lemma 7.4, we obtain

$$v_{1k} \geq (2K + 1)M_1 = M, \quad k \in \mathbb{N}_1,$$

and again (7.24) follows from (7.23).

Define the sequence of intervals $\delta_k = [s_k, t_k^1]$, $k \in \mathbb{N}_1$, in the following way. Choose $s_k \in (t_{k-1}^0, t_k^1)$ such that $|y(s_k)| = \frac{v_{0k}}{2}$, $k \geq k_0$. Then, from (7.24) and from the properties of Type I solutions,

$$|y(t)| \geq \frac{v_{0k}}{2} \geq K \text{ on } \delta_k, \ k \geq k_0. \tag{7.25}$$

Let $\Delta_k = t_k^1 - s_k$. Using (7.22) and (7.25), we obtain

$$|y(t_k^1)| - |y(s_k)| \leq \frac{v_{0k}}{2} = \int_{\delta_k} |y'(t)| dt = \int_{\delta_k} a_1 |y^{[1]}|$$

$$\leq \max_{s \in \mathbb{R}_+} a_1(s) \Delta_k v_{1k} \leq \max_{s \in \mathbb{R}_+} a_1(s) \Delta_k M_1 (1 + v_{0k})$$

for $k \geq k_0$. Since $a_1$ is bounded from above and (7.25) holds, there exists $\gamma > 0$ such that

$$\Delta_k \geq \gamma > 0, \ k \geq k_0.$$

Finally, applying (7.4), we have

$$\int_0^\infty y(t) f(y(t), \dots, y^{[3]}(t)) dt \geq \sum_{k=k_0}^\infty \int_{\Delta_k} y f(y, \dots, y^{[3]}) dt \geq \sum_{k=k_0}^\infty \int_{\Delta_k} dt = \infty.$$

We have thus proved that this oscillatory solution is of the nonlinear limit–point type.

Let $y$ be a Type II solution. From (7.19) we obtain

$$|y(t) - y(\tau)| = \int_\tau^t |y'(s)| ds = \int_\tau^t a_1 |y^{[1]}| \geq |y^{[1]}(\tau)| \int_\tau^t a_1(s) ds \to \infty$$

for $t \to \infty$. Hence, $\lim_{t \to \infty} |y(t)| = \infty$. From (7.4),

$$\int_\tau^t y(t) f(y(t), \dots, y^{[3]}(t)) dt \geq \int_{t_1}^\infty y(t) f(y(t), \dots, y^{[3]}(t)) dt \geq \int_{t_1}^\infty dt = \infty,$$

where $t_1 \geq \tau$, $t_1 \geq \sigma$ is such that $|y(t_1)| \geq K$. Thus, the nonoscillatory solution with initial conditions (7.20)–(7.21) is of the nonlinear limit–point type.  □

**Remark 7.2.** From the proof of Theorem 7.2, it follows that *every* Type II solution is of the nonlinear limit–point type.

## 7.2. Sublinear Equations in Self-Adjoint Form

Consider the fourth-order differential equation of the form

$$(p_2(t)y'')'' - (p_1(t)y')' + p_0(t)f(y) = 0, \tag{7.26}$$

where $p_i \in C^i(\mathbb{R}_+)$, $i = 0, 1, 2$, are real-valued functions, $p_2 > 0$, $f : \mathbb{R} \to \mathbb{R}$ is continuous, $yf(y) \geq 0$ on $\mathbb{R}$, and there exist $K > 0$ and $C > 0$ such that

$$\frac{1}{|x|} \leq |f(x)| \leq C|x| \quad \text{for } |x| \geq K. \tag{7.27}$$

For equation (7.26), the definitions of nonlinear limit–point and limit–circle take the same form as for equation (7.1), i.e., (7.26) is of the *nonlinear limit–circle type* if every continuable solution $y$ satisfies

$$\int_0^\infty y(t) f(y(t)) dt < \infty,$$

and if there is at least one continuable solution $y$ such that

$$\int_0^\infty y(t) f(y(t)) dt = \infty,$$

then equation (7.26) is of the *nonlinear limit–point type*.

Suppose that the second order differential equation

$$(p_2(t)u')' - p_1(t)u = 0 \tag{7.28}$$

is nonoscillatory, and let $h > 0$ be a solution of (7.28). Then, similar to what we did for third order equations (see Section 6.3), we have

$$(p_2(t)u')' - p_1(t)u = \frac{1}{h}\left(p_2 h^2 \left(\frac{u}{h}\right)'\right)',$$

and so letting $u = y'$, we obtain

$$(p_2(t)y'')'' - (p_1(t)y')' = ((p_2(t)u')' - p_1(t)u)' = \left(\frac{1}{h}\left(p_2 h^2 \left(\frac{y'}{h}\right)'\right)'\right)'.$$

Hence, equation (7.26) can be written in the form

$$\left(\frac{1}{h(t)}\left(p_2(t)h^2(t)\left(\frac{y'}{h(t)}\right)'\right)'\right)' = -p_0(t)f(y), \qquad (7.29)$$

which is a special case of equation (7.1).

Applying Theorem 7.1 to equation (7.29), we obtain the following result.

**Theorem 7.3.** *Assume that $p_0 \le 0$, (7.27) holds, and equation (7.28) is nonoscillatory. Then equation (7.26) is of the nonlinear limit–point type.*

To apply Theorem 7.2, consider the special case of equation (7.26), namely,

$$y^{(4)} - (p_1(t)y')' + p_0(t)f(y) = 0, \qquad (7.30)$$

where $f$ satisfies (7.27). We will need the following auxiliary results on the behavior of solutions of equation (7.28).

**Proposition 7.1.** *([93, Theorems 1, 3 and 4']) Let $p_1(t) \ge 0$ and not identically zero on an infinite interval. Then:*

(i) *There exists a positive nonincreasing solution of equation (7.28).*

(ii) *Every positive nonincreasing solution of equation (7.28) tends to zero if and only if*

$$I(p_1, p_2) = \int_0^\infty |p_1(s)| \int_0^s \frac{1}{p_2(u)}\, du\, ds = \infty.$$

**Proposition 7.2.** *([26, Theorem 1]) Let $p_1(t) < 0$ and $I(p_1, p_2) < \infty$. Then equation (7.28) has a positive bounded nondecreasing solution.*

**Lemma 7.6.** *Let (7.27) hold, $p_0 \ge 0$ be bounded from above, and let the equation*

$$u'' - p_1 u = 0 \qquad (7.31)$$

*be nonoscillatory. If this equation has either a nonincreasing solution $h$ such that $\lim_{t\to\infty} h(t) > 0$ or a nondecreasing solution $h$ such that $\lim_{t\to\infty} h(t) < \infty$, then equation (7.30) is of the nonlinear limit–point type.*

*Proof.* We rewrite equation (7.30) in the form (7.29) with $p_2 = 1$, and then we apply Theorem 7.2.                                                                    □

Using Lemma 7.6 and Propositions 7.1 and 7.2, we obtain the following result.

**Theorem 7.4.** *Let $p_0 \geq 0$ be bounded from above, $p_1 \neq 0$, (7.27) hold, and*

$$\int_0^\infty u|p_1(u)|du < \infty.$$

*Then equation (7.30) is of the nonlinear limit–point type.*

*Proof.* The right hand inequality in (7.27) guarantees that there are no singular solutions (see Corollary 2.1 above). Suppose $p_1 > 0$; then equation (7.31) is nonoscillatory. By Proposition 7.1, there exists a positive nonincreasing solution of equation (7.31) tending to a nonzero constant. The conclusion then follows from Lemma 7.6.

Similarly, if $p_1 < 0$, we use Proposition 7.2 and Lemma 7.6.                      □

## 7.3. Two-Term Equations

In this section, we consider the nonlinear fourth order differential equation

$$y^{(4)} = r(t)f(y, y', y'', y'''), \tag{7.32}$$

where $r \in L_{\text{loc}}(\mathbb{R}_+)$, and often assume that

$$r \text{ does not change sign on } [t_0, \infty), \ t_0 \geq 0, \tag{7.33}$$

$f : \mathbb{R}^4 \to \mathbb{R}$ is continuous, and

$$x_1 f(x_1, x_2, x_3, x_4) \geq 0 \text{ on } \mathbb{R}^4. \tag{7.34}$$

Our nonlinear limit–point result for equation (7.32) is contained in the following theorem.

**Theorem 7.5.** *Suppose (7.33)–(7.34) hold and there exist positive constants $M_i$, $i = 1, 2$ such that*

$$M_1|x_1| \leq |f(x_1, x_2, x_3, x_4)| \leq M_2(1 + |x_1|) \text{ on } \mathbb{R}^4. \tag{7.35}$$

*Then (7.32) is of the nonlinear limit–point type.*

To prove this theorem, we need the following preliminary results, the first of which is due to Kiguradze and Chanturia [80].

**Lemma 7.7.** *Let* $y \in C^3[\tau, \infty)$, $\tau \in (-\infty, \infty)$,

$$y^{(i)}(t)y(t) > 0 \text{ for } i = 0, 1, 2, \text{ and } y'''(t)y(t) \geq 0 \text{ for } t \in [\tau, \infty).$$

*Then,*

$$|y(t)y'(t)y''(t)|^{-\frac{1}{12}} \geq \gamma \int_t^{\infty} |y'''(s)|^{\frac{1}{4}} |y(s)|^{-\frac{1}{2}} \, ds, \qquad (7.36)$$

*where* $\gamma$ *is a suitable positive constant.*

*Proof.* See Lemma 11.2 in [80] with $n = 3$, $\alpha = \frac{1}{4}$, and $k = 1$. $\qquad \square$

Let $W(\mathbb{R}_+)$ be the subspace of $L^2(\mathbb{R}_+)$ consisting of those functions $y \in C^1(\mathbb{R}_+)$ such that $y''$ is absolutely continuous on $\mathbb{R}_+$ and $y''' \in L^2(\mathbb{R}_+)$. The next result is a special case of Theorem 1.4 in Kwong and Zettl [90] and its general form will be presented in Lemma 8.1.

**Lemma 7.8.** *There exists a constant* $M > 0$ *such that*

$$\int_0^{\infty} [y'(s)]^2 \, ds \leq M \left[ \int_0^{\infty} y^2(s) \, ds \int_0^{\infty} (y''(s))^2 \, ds \right]^{\frac{1}{2}} \qquad (7.37)$$

*for all* $y \in W(\mathbb{R}_+)$.

A straightforward computation verifies the following lemma.

**Lemma 7.9.** *Let* $r(t) \leq 0$ *on* $\mathbb{R}_+$ *and* $y$ *be a solution of* (7.32). *Then the function*

$$z(t; y) = z(t) = -\frac{1}{2}y^2(t) + 2 \int_0^t \int_0^s [y'(u)]^2 \, du \, ds$$

*satisfies*

$$z'(t) = -yy' + 2 \int_0^t [y'(s)]^2 \, ds,$$
$$z''(t) = -yy'' + y'^2, \quad z'''(t) = -yy''' + y'y'', \qquad (7.38)$$

*and*

$$z^{(4)}(t) = -yy^{(4)} + y''^2 \geq 0. \qquad (7.39)$$

The next lemma classifies certain types of solutions of (7.32).

**Lemma 7.10.** *Suppose $r(t) \leq 0$ on $\mathbb{R}_+$. Let $y : \mathbb{R}_+ \to \mathbb{R}$ be a solution of (7.32) with $y^{(i)}(0) > 0$ for $i = 0, 1, 2, 3$, and such that the $z$ in Lemma 7.9 satisfies $z'''(0) > 0$ for this solution. Then either $y$ is oscillatory or there exists $\tau \in \mathbb{R}_+$ such that*

$$yy^{(i)} > 0 \quad on \quad [\tau, \infty), \quad i = 0, 1, 2, \quad and \quad yy''' \geq 0 \quad on \quad [\tau, \infty). \quad (7.40)$$

*Proof.* The conclusion follows immediately from [5, Lemma 2].                     □

*Proof of Theorem 7.5.* If $r(t) \geq 0$, the conclusion follows from Theorem 7.1. Suppose $r(t) \leq 0$ on $\mathbb{R}_+$. Consider a solution $y$ of (7.32) having the initial conditions

$$y(0) = 0, \quad y'(0) = y''(0) = y'''(0) = 1.$$

Then, the function $z$ defined in Lemma 7.9 satisfies

$$z(0) = z'(0) = 0, \quad z''(0) = z'''(0) = 1. \quad (7.41)$$

In view of this, and the monotonicity of $z'''$, we obtain that

$$z^{(j)} > 0 \text{ are nondecreasing for } j = 0, 1, 2 \text{ and all } t > 0. \quad (7.42)$$

It follows from the second inequality in (7.35) that $y$ is defined on all of $\mathbb{R}_+$ (see Corollary 2.1). Since $y^{(i)}(t) > 0, i = 0, 1, 2, 3$, in some right-hand neighborhood of $t = 0$, we have that $y$ is either oscillatory or satisfies (7.40).

If $y$ satisfies (7.40), then condition (7.35) implies

$$\int_0^\infty y(t) f(y(t), y'(t), y''(t), y'''(t)) \, dt \geq M_1 \int_0^\infty y^2(t) \, dt$$

$$\geq M_1 \int_\tau^\infty y^2(t) \, dt \geq M_1 y^2(\tau) \int_\tau^\infty dt = \infty, \quad (7.43)$$

and so $y$ is a nonlinear limit–point type solution.

Suppose $y$ is oscillatory. If $\int_0^\infty y^2(t) \, dt = \infty$, the conclusion follows from the first inequality in (7.43), so suppose that

$$\int_0^\infty y^2(t) \, dt = K < \infty. \quad (7.44)$$

For each $s \in (0, \infty)$ there exists a function $Y_s : \mathbb{R}_+ \to \mathbb{R}$ such that

$$Y_s(t) = \begin{cases} y(t) & \text{for } t \in [0, s], \\ 0 & \text{for } t \in [s + 1, \infty), \end{cases}$$

$Y_s^{(j)}$ are continuous for $j = 0, 1, 2$, on $(s, s + 1)$, $Y_s'''$ is absolutely continuous on $(s, s + 1)$, and

$$\int_s^{s+1} Y_s^2(u) \, du \le 1, \quad \int_s^{s+1} [Y_s''(u)]^2 \, du \le 1.$$

Then, it follows from (7.41) and Lemma 7.8 that

$$z(s) \le 2s \int_0^s [y'(u)]^2 \, du \le 2s \int_0^\infty [Y_s'(u)]^2 \, du$$

$$\le 2sM \left[ \int_0^\infty Y_s^2(u) \, du \int_0^\infty [Y_s(u)'']^2 \right]^{\frac{1}{2}} du$$

$$\le 2Ms \left[ \left( \int_0^s y^2(u) \, du + 1 \right) \left( \int_0^s [y''(u)]^2 \, du + 1 \right) \right]^{\frac{1}{2}}.$$

From this, (7.39), (7.41), and (7.44), we have

$$z(s) \le 2M(K + 1)^{\frac{1}{2}} s \left( \int_0^s z^{(4)}(u) \, du + 1 \right)^{\frac{1}{2}}$$

$$\le K_1 s (z'''(s) - z'''(0) + 1)^{\frac{1}{2}} \le K_1 s (z'''(s))^{\frac{1}{2}},$$

where $K_1 = 2M(K + 1)^{\frac{1}{2}}$. Thus,

$$z'''(s) \ge \frac{z^2(s)}{K_1^2 s^2}, \quad s > 0. \tag{7.45}$$

We will show that the inequality (7.45) has no positive nondecreasing solutions on $\mathbb{R}_+$. It follows from (7.39) and (7.41) that

$$z(t) = z(t) - z(0) = \int_0^t z'(s) \, ds \le z'(t)t, \quad t > 0,$$

so

$$z'(t) \ge \frac{z(t)}{t}, \quad t > 0. \tag{7.46}$$

Similarly,

$$z''(t) \ge \frac{z'(t)}{t} \ge \frac{z(t)}{t^2}, \quad t > 0. \tag{7.47}$$

In view of (7.42), we can apply Lemma 7.7 to $z$, so (7.46) and (7.47) yield

$$\left( \frac{t}{z(t)} \right)^{\frac{1}{4}} \ge \left( \frac{1}{z(t)z'(t)z''(t)} \right)^{\frac{1}{12}} \ge \gamma \int_t^\infty [z'''(s)]^{\frac{1}{4}} [z(s)]^{-\frac{1}{2}} \, ds$$

$$\ge \gamma \int_t^\infty \frac{ds}{(K_1 s)^{\frac{1}{2}}} \, ds = \infty$$

for $t > 0$. Thus, $z \equiv 0$ on $\mathbb{R}_+$ and this contradicts (7.42), so (7.44) cannot hold.                                                                                                □

**Corollary 7.1.** *The equation*

$$y^{(4)} = r(t)y,$$

*where $r(t)$ does not change its sign and is bounded from below, is always of the nonlinear limit–point type.*

**Remark 7.3.** Eastham and Grudniewicz [41, Theorem 1] proved Corollary 7.1 for $r(t) \geq 0$. For $r(t) \leq 0$, this result appears to be new.

## 7.4. Linear Equations

In this section, we consider the linear fourth order differential equation

$$Ly \equiv y^{(4)} + p(t)y'' + q(t)y' + r(t)y = 0, \tag{7.48}$$

where $p, q, r$ are real continuous functions on $\mathbb{R}_+$. Equation (7.48) is more general than the linear form of equations (7.1) and (7.26) since we do not assume it can be expressed with quasiderivatives nor do we assume that it is in self-adjoint form. However, we will still say that equation (7.48) is (linear) limit–circle if all solutions belong to $L^2$.

The following lemma is immediate.

**Lemma 7.11.** *Suppose that*

$$r'(t) \geq 0 \text{ and } p'(t) - 2q(t) \geq 0 \quad \text{for } t \geq 0, \tag{7.49}$$

*and let $y$ be a solution of (7.48). Then the function $F$ defined by*

$$F(t; y) = F(t) = ry^2 + py'^2 - y''^2 + 2y'y'''$$

*satisfies*

$$F'(t) = r'y^2 + (p' - 2q)(y')^2 \geq 0.$$

**Theorem 7.6.** *Suppose $r \leq 0$ on $\mathbb{R}_+$. If either*

    *(i)  (7.49) holds, or*

    *(ii) $p(t) \leq 0$ and $q(t) \leq 0$ for $t \geq 0$,*

*then (7.48) is not of the limit–circle type, i.e., there is at least one solution that does not belong to $L^2$.*

*Proof.* (i) Let $y$ be a solution of (15) with the initial conditions

$$y(0) = y'(0) = 1, \quad y''(0) = 0, \quad y'''(0) > -\frac{r(0) + p(0)}{2}.$$

Then, the function $F$ from Lemma 7.11 satisfies

$$F(0, y) > 0. \tag{7.50}$$

We will show that

$$y'(t) > 0 \quad \text{for } t \geq 0.$$

Suppose, to the contrary, that there exists $T \in (0, \infty)$ such that $y'(T) = 0$. Then

$$F(T; y) = r(T)y^2(T) - y''^2(T) \leq 0,$$

which contradicts (7.50) and the monotonicity of $F$. Thus, $y$ is nondecreasing and positive on $\mathbb{R}_+$, so

$$\int_0^\infty y^2(t)\, dt \geq y^2(0) \int_0^\infty dt = \infty,$$

i.e., (7.48) is not limit–circle.

(ii) Let $y$ be a solution of (7.48) with the initial conditions $y^{(i)}(0) > 0$ for $i = 0$, 1, 2, 3. Then, it follows from (7.48) that $y^{(4)}(t) > 0$ in a right-hand neighborhood of 0 as long as $y^{(i)}(t) > 0$ for $i = 0, 1, 2, 3$. Suppose there exist $j \in \{0, 1, 2, 3\}$ and $T > 0$ such that $y^{(j)}(T) = 0$ and $y^{(i)}(t) > 0$ for $t \in [0, T)$ and $i = 0, 1, 2, 3$. Then, there exists $\tau \in [0, T)$ such that $y^{(j+1)}(\tau) < 0$, which is a contradiction. Hence, $y$ is positive and nondecreasing on $\mathbb{R}_+$, and the conclusion follows immediately. $\square$

A straightforward calculation proves the following lemma.

**Lemma 7.12.** *Let $y$ be a solution of (7.48) and $z(t) = \left[1 + \left(\frac{q(t)}{r(t)}\right)'\right]^{-1}$. Suppose that for $t \geq 0$,*

$$A(t) = 3z'(t)r(t) - z(t)r'(t) \geq 0 \tag{7.51}$$

*and*

$$B(t) = \frac{2q(t)z(t)}{r(t)} - \left(\frac{z'(t)}{r(t)}\right)'' + \left(\frac{p(t)z(t)}{r(t)}\right)' - \frac{2p(t)z'(t)}{r(t)} \geq 0. \tag{7.52}$$

*Then the function F defined by*

$$F(t; y) = F(t) = \frac{2z'}{r} y''y''' + \frac{z}{r}(y''')^2 + 2zyy''$$

$$+ \frac{2qz}{r} y'y'' - \left( \left( \frac{z'}{r} \right)' - \frac{pz}{r} \right)(y'')^2 - (y')^2$$

*satisfies*

$$F'(t) = B(t)(y'')^2 + \frac{A(t)}{r^2(t)}(y''')^2 \geq 0.$$

**Theorem 7.7.** *Let $z$ be defined as in Lemma 7.12, let (7.51) and (7.52) hold, and suppose that $\frac{z(t)}{r(t)} < 0$ for $t \geq 0$. Then (7.48) is not of the limit–circle type.*

*Proof.* We proceed in a manner similar to the proof of Theorem 7.6 . Let $y$ be a solution of (7.48) with initial conditions $y^{(i)}(0) > 0$, $i = 0, 1, 2$, and such that the function $F$ given in Lemma 7.12 satisfies $F(0, y) > 0$. We will show that $y''(t) > 0$ for $t \geq 0$. Suppose, to the contrary, that there exists $T \in (0, \infty)$ such that $y''(T) = 0$. Then,

$$F(T; y) = \frac{z(T)}{r(T)}(y'''(T))^2 - (y'(T))^2 \leq 0,$$

which contradicts $F(0) > 0$ and the monotonicity of $F$. Thus, $y'$ is nondecreasing and positive on $\mathbb{R}_+$. The conclusion of the theorem is immediate. □

**Open Problems.**

**Problem 7.1.** *What conditions can be placed on the coefficients in equations (7.1), (7.26), or (7.30) so that a limit–circle type solution is oscillatory?*

**Problem 7.2.** *In the case of second order equations, the limit–circle property is not enough to guarantee that all solutions are bounded. What is the situation for fourth order equations?*

**Problem 7.3.** *What conditions on the coefficients in equations (7.1), (7.26), or (7.30) can be imposed so that all proper solutions tend to zero as t tends to infinity?*

**Notes.** Theorems 7.1–7.4 are based on results of Bartušek, Došlá, and Graef [13] while Theorems 7.5–7.7 are taken from Bartušek, Došlá, and Graef [12].

# Chapter 8

# Nonlinear Differential Equations of $n$-th Order

In this chapter, we obtain some nonlinear limit–point results for equations of order $n \geq 2$. At the end of the chapter, the results are applied to the sublinear Thomas–Fermi equation.

## 8.1. Introduction

Here, we consider the $n$-th order nonlinear differential equation

$$y^{(n)} = r(t)f(y, y', \ldots, y^{(n-1)}) \tag{8.1}$$

where $n \geq 2$, $r \in L_{\text{loc}}(\mathbb{R}_+)$,

$$r \text{ does not change sign on } \mathbb{R}_+, \tag{8.2}$$

$f : \mathbb{R}^n \to \mathbb{R}$ is continuous, and

$$x_1 f(x_1, \ldots, x_n) \geq 0 \quad \text{on} \quad \mathbb{R}^n. \tag{8.3}$$

As for third and fourth order equations, in the case $r \geq 0$ there exists a limit–point type solution under weak assumptions on the nonlinearity $f$, as the following result shows.

**Theorem 8.1.** *Suppose* (8.3) *holds,* $r \geq 0$ *on* $\mathbb{R}_+$, *and there exist constants* $M > 0$ *and* $M_1 > 0$ *such that*

$$\frac{1}{|x_1|} \leq |f(x_1, \ldots, x_n)| \leq M_1|x_1| \tag{8.4}$$

*for $x_1 \geq M$ or $x_1 \leq -M$, $x_i \in \mathbb{R}$, $i = 2, \ldots, n$. Then equation (8.1) is of the nonlinear limit–point type.*

*Proof.* Let $y$ be a solution of (8.1) satisfying the initial conditions

$$y(0) \geq M, \quad y^{(i)}(0) > 0, \quad i = 1, 2, \ldots, n - 1.$$

Then, $y$ is defined on $\mathbb{R}_+$ by Corollary 2.1. Since $r \geq 0$, we see that $y^{(n)}(t) \, y(t) \geq 0$ on $\mathbb{R}_+$, so

$$y(t) \geq M \text{ and } y \text{ is nondecreasing on } \mathbb{R}_+.$$

Thus,

$$\int_0^\infty y(t) f(y(t), \ldots, y^{(n-1)}(t)) \, dt \geq \int_0^\infty dt = \infty,$$

and so $y$ is of the nonlinear limit-point type. If (8.4) holds for $x_1 \leq -M$, the proof is similar.                                                                              □

**Remark 8.1.** For $n = 3$, Theorem 8.1 was proved under a weaker nonlinear condition on $f$ in Theorem 6.1. Theorem 3.14 above is also a special case of Theorem 8.1. For a fourth order version of Theorem 8.1 for equations with quasiderivatives see Theorem 7.1 above.

In the remainder of this chapter, we only consider the case $r(t) \leq 0$. In the study of asymptotic properties of solutions of $n$-th order equations, the order itself plays an important role. Hence, we divide the positive integers into the following three disjoint sets:

$$n = 4k,$$
$$n = 2k + 1,$$
$$n = 4k + 2,$$

where $k = 0, 1, 2, \ldots$. We will use the usual notation $[\![i]\!]$ to denote the greatest integer function of $i$.

## 8.2. Basic Lemmas

Limit–point properties of equation (8.1) in the case $r \leq 0$ are more interesting. To pursue this direction of study, we will need some auxiliary results. Our first three lemmas provide some useful integral inequalities. Let $W^m(\mathbb{R}_+)$ be the subspace of $L^2(\mathbb{R}_+)$ consisting of those functions $y \in L^2(\mathbb{R}_+)$ with $y^{(m)} \in L^2(\mathbb{R}_+)$.

**Lemma 8.1.** ([90, Theorem 1.4]) *Let $j$ and $m$ be integers with $0 \leq j < m$, $1 \leq p < \infty$, $1 \leq q < \infty$, and $1 \leq w < \infty$. Then, there exists a constant $M > 0$ such that*

$$\left( \int_0^\infty [y^{(j)}(s)]^q \, ds \right)^{1/q} \leq M \left[ \int_0^\infty y^p(s) \, ds \right]^{\frac{\alpha}{p}} \left[ \int_0^\infty [y^{(m)}(s)]^w \, ds \right]^{\frac{1-\alpha}{w}}$$

*for all $y \in W^m(\mathbb{R}_+)$ if and only if $\frac{m}{q} \leq \frac{m-j}{p} + \frac{j}{w}$ where*

$$\alpha = \frac{m - j - \frac{1}{w} + \frac{1}{q}}{m - \frac{1}{w} + \frac{1}{p}}.$$

We let $\tilde{C}^m(\mathbb{R}_+)$, $m \geq 0$, denote the subspace of $C^m(\mathbb{R}_+)$ consisting of those functions $y$ such that $y^{(m)}$ is absolutely continuous on every compact subinterval of $\mathbb{R}_+$.

**Lemma 8.2.** ([80, Lemma 11.2] with $k = 1$) *Let $m \geq 2$, $y \in \tilde{C}^{m-1}[\tau, \infty)$, $\tau \in (-\infty, \infty)$, $y^{(i)}(t)y(t) > 0$ for $i = 0, 1, \ldots, m - 1$, and $y^{(m)}(t)y(t) \geq 0$ for $t \in [\tau, \infty)$. Then, for any $\alpha \in (0, \frac{1}{m})$, we have*

$$\left( \prod_{i=0}^{m-1} |y^{(i)}(t)| \right)^{-\varepsilon} \geq \lambda \int_t^\infty |y^{(m)}(s)|^\alpha |y(s)|^\beta \, ds, \quad t > \tau,$$

*where $\lambda$ is a suitable positive constant, $\varepsilon = \frac{2(1-m\alpha)}{m(m-1)}$, and $\beta = \frac{(m+1)\alpha-2}{m-1}$.*

**Lemma 8.3.** *Let $y \in \tilde{C}^0(\alpha, \beta)$, $-\infty < \alpha < \beta < \infty$, $y' \in L^2(\alpha, \beta)$, and $y(t_0) = 0$, where $t_0 \in [\alpha, \beta]$. Then,*

$$\int_\alpha^\beta y^2(t) \, dt \leq \frac{\pi^2}{4} (\beta - \alpha)^2 \int_\alpha^\beta [y'(t)]^2 \, dt.$$

*Proof.* See, for example, [79, Lemma 4.7]. □

**Notation 8.1.** For $v \in C^0(\mathbb{R}_+)$, set

$$J_0(t; v) = v(t) \quad \text{and} \quad J_m(t; v) = \int_0^t J_{m-1}(s; v(s)) \, ds, \quad m = 1, 2, \ldots.$$

Our next lemma introduces an energy type function that will play a crucial role in what follows.

**Lemma 8.4.** *Let* $n_0 = [\![\frac{n}{2}]\!]$ *and suppose that (8.3) holds and* $r(t) \leq 0$ *on* $\mathbb{R}_+$. *Let* $y$ *be an oscillatory solution of (8.1) and define the function* $z$ *by*

$$z(t; y) = z(t) = \sum_{v=0}^{n-n_0-1} (-1)^{v+1} a_v J_{2v}(t; [y^{(v)}]^2),$$

*where*

$$a_v = \binom{n-v}{v} \frac{n}{2(n-v)} > 0.$$

*(a) If* $n = 2k + 1$, *then*

$$z^{(n-1)}(t) = \sum_{v=0}^{n_0-1} (-1)^{v+1} y^{(n-v-1)}(t) y^{(v)}(t) + \frac{1}{2}(-1)^{n_0+1} [y^{(n_0)}(t)]^2$$

*and*

$$z^{(n)}(t) = -y(t) y^{(n)}(t).$$

*In addition, if there exist positive constants* $M_i$, $i = 1, 2, 3$, *and a continuous function* $g : \mathbb{R}_+ \to \mathbb{R}$ *such that* $M_1 \leq |r(t)| \leq M_2$, $g(0) = 0$, $g(x) > 0$ *for* $x > 0$, $\liminf_{x \to \infty} g(x) > 0$, *and*

$$g(|x_1|) \leq |f(x_1, \ldots, x_n)| \leq M_3(1 + |x_1|) \quad on \quad \mathbb{R}^n,$$

*then* $\lim_{t \to \infty} z^{(n-1)}(t) \in \{0, \infty\}$.

*(b) If* $n = 4k$, *then*

$$z^{(i)}(t) = \sum_{v=0}^{[\![\frac{i}{2}]\!]} (-1)^{v+1} c_{vi} y^{(v)}(t) y^{(i-v)}(t) + \sum_{v=[\![\frac{i}{2}]\!]+1}^{n_0-1} (-1)^{v+1} c_{vi} J_{2v-i}(t; [y^{(v)}]^2)$$

*for* $i = 0, 1, \ldots, n - 3$,

$$z^{(n-2)}(t) = \sum_{v=0}^{n_0-2} (-1)^{v+1} (v + 1) y^{(v)}(t) y^{(n-2-v)}(t) + \frac{n_0}{2} [y^{(n_0-1)}(t)]^2,$$

$$z^{(n-1)}(t) = \sum_{v=0}^{n_0-1} (-1)^{v+1} y^{(n-v-1)}(t) y^{(v)}(t),$$

*and*

$$z^{(n)}(t) = -y^{(n)}(t) y(t) + [y^{(n_0)}(t)]^2 \geq 0,$$

*for* $t \in \mathbb{R}_+$, *where* $c_{vi}$ *are constants. Moreover,* $\lim_{t \to \infty} z^{(n-1)}(t) \in \{0, \infty\}$.

*Proof.* This lemma is proved in [2, Lemma 3.1, Lemma 3.4, and Lemma 3.9] except for the explicit expressions for $z^{(i)}$, $i = 1, \ldots, n - 2$, in case (b). This can be proved by a direct computation noting that the coefficients $c_{vi}$ are given by

$$c_{v0} = a_v, \quad v = 0, 1, \ldots, n_0 - 2, \quad c_{0i} = a_0, \quad i = 1, \ldots, n - 3,$$

$$c_{v,i+1} = \begin{cases} c_{vi} - c_{v-i,i} & \text{for } v = 1, \ldots, [\![\frac{i-1}{2}]\!], \\ \sigma c_{vi} - c_{v-1,i} & \text{for } v = [\![\frac{i+1}{2}]\!], \\ c_{vi} & \text{for } v \geq [\![\frac{i+1}{2}]\!] + 1, \end{cases}$$

where $\sigma = 2$ if $i$ is even and $\sigma = 1$ if $i$ is odd. $\qquad\square$

The next three lemmas describe some asymptotic properties of solutions of equation (8.1).

**Lemma 8.5.** *Suppose* $r(t) \leq 0$ *on* $\mathbb{R}_+$, *and let* $y : \mathbb{R}_+ \to \mathbb{R}$ *be a solution of* (8.1) *with* $y^{(i)}(0) > 0$ *for* $i = 0, 1, \ldots, n - 1$. *Then* $y$ *is one of the following two types.*

*Type I. There exists* $\tau \in \mathbb{R}_+$ *such that* $yy^{(i)} > 0$ *on the interval* $[\tau, \infty)$ *for* $i = 0, 1, \ldots, n - 2$ *and* $yy^{(n-1)} \geq 0$ *on* $[\tau, \infty)$.

*Type II.* $y$ *is an oscillatory solution.*

*Proof.* Suppose that $y$ is a nonoscillatory solution. According to [5, Lemmas 2 and 9], there exist $s \in \{0, 1, \ldots, n - 1\}$ and $\tau \in \mathbb{R}_+$ such that

$$y^{(s)}(t) \, y^{(j)}(t) \begin{cases} \geq 0 & \text{for} \quad j \in \{0, 1, \ldots, s\}, \\ \leq 0 & \text{for} \quad j \in \{s + 1, \ldots, n - 1\}, \end{cases}$$

and

$$\prod_{i=0}^{n-2} y^{(i)}(t) \neq 0 \quad \text{on} \quad [\tau, \infty).$$

Then, $y$ is defined on $\mathbb{R}_+$, i.e., the only possible case is that $s = n - 1$. $\qquad\square$

**Lemma 8.6.** ([6, Theorem 2]) *Suppose* $n \geq 4$ *and* $r(t) \leq 0$ *on* $\mathbb{R}_+$, *and let* $y : \mathbb{R}_+ \to \mathbb{R}$ *be an oscillatory solution of* (8.1) *with* $y^{(i)}(0) > 0$ *for* $i = 0, 1, \ldots, n-1$. *Then,* $y^{(j)}$, $j = 1, 2, \ldots, n - 3$, *are unbounded.*

**Lemma 8.7.** ([20, §V.2. Theorem 1]) *Let* $y : \mathbb{R}_+ \to \mathbb{R}$ *be such that* $y \in L^p(\mathbb{R}_+)$ *and* $y' \in L^p(\mathbb{R}_+)$ *where* $p \geq 1$. *Then* $y$ *is bounded on* $\mathbb{R}_+$.

## 8.3. Limit–Point Results

In this section, we present some limit–point results for equation (8.1) with $r \leq 0$.

**Theorem 8.2.** *Let $n = 4k$. Suppose (8.3) holds, $r \leq 0$ on $\mathbb{R}_+$, and there exist positive constants $M_i$, $i = 1, 2$, and $\lambda \in (0, 1]$ such that*

$$M_1 |x_1|^\lambda \leq |f(x_1, x_2, \ldots, x_n)| \leq M_2(1 + |x_1|) \quad \text{on} \quad \mathbb{R}^n. \tag{8.5}$$

*Then equation (8.1) is of the nonlinear limit–point type.*

*Proof.* Consider a solution $y$ of (8.1) having the initial conditions

$$y^{(i)}(0) = 0, \quad i = 0, 1, \ldots, n_0 - 2, \quad y^{(j)}(0) = 1, \quad j = n_0 - 1, \ldots, n - 1. \tag{8.6}$$

It follows from the second inequality in (8.5) that $y$ is defined on $\mathbb{R}_+$ (see Chapter 2). According to (8.6), $y^{(i)}(t) > 0$ for $i = 0, 1, \ldots, n - 1$, in some right-hand neighborhood of $t = 0$, and so $y$ is either a Type I or Type II solution as defined in Lemma 8.5.

   (i) Let $y$ be of Type I. Then by condition (8.5),

$$\int_0^\infty y(t) f(y(t), \ldots, y^{(n-1)}(t)) \, dt \geq M_1 \int_0^\infty |y(t)|^{1+\lambda} \, dt \geq M_1 \int_\tau^\infty |y(t)|^{1+\lambda} dt$$

$$\geq M_1 |y(\tau)|^{1+\lambda} \int_\tau^\infty dt = \infty. \tag{8.7}$$

Hence, $y$ is a nonlinear limit–point type solution.

   (ii) Now, let $y$ be of Type II. For convenience, let

$$\mu = 1 + \lambda \quad \text{and} \quad \delta = n\mu - \mu + 2.$$

If $\int_0^\infty |y(t)|^\mu \, dt = \infty$, the conclusion follows from the first inequality in (8.5), so suppose that

$$\int_0^\infty |y(t)|^\mu \, dt = K < \infty. \tag{8.8}$$

For each $s \in (0, \infty)$, there exists a function $Y_s : \mathbb{R}_+ \to \mathbb{R}$ such that

$$
\begin{cases}
Y_s(t) = \begin{cases} y(t) & \text{for } t \in [0, s], \\ 0 & \text{for } t \in [s + 1, \infty), \end{cases} \\
Y_s^{(j)} \text{ are continuous for } j = 0, 1, \ldots, [\![\frac{n}{2}]\!] \text{ on } [s, s + 1], \\
\int_s^{s+1} |Y_s(u)|^\mu \, du \leq 1, \text{ and } \int_s^{s+1} [Y_s^{(n_0)}(u)]^2 \, du \leq 1.
\end{cases} \tag{8.9}
$$

Now consider the function $z$ defined in Lemma 8.4. Clearly, $z$ is defined on $\mathbb{R}_+$. Thus, in view of (8.6) and the monotonicity of $z^{(n-1)}$, we have

$$z^{(i)}(0) \geq 0, \quad i = 0, 1, \ldots, n - 2, \quad z^{(n-1)}(0) = 1, \tag{8.10}$$

$$z^{(j)}(t) > 0 \text{ are nondecreasing for } j = 0, 1, \ldots, n - 1 \text{ and all } t > 0. \tag{8.11}$$

Then, by (8.6) and Lemmas 8.3 and 8.4,

$$
\begin{aligned}
z(s) &\leq \sum_{\nu=0}^{n_0-1} a_\nu s^{2\nu-1} \int_0^s [y^{(\nu)}(u)]^2 \, du \\
&\leq \sum_{\nu=0}^{n_0-1} a_\nu \left(\frac{\pi}{2}\right)^{2(n_0-\nu-1)} s^{2n_0-3} \int_0^s [y^{(n_0-1)}(u)]^2 \, du \\
&= C s^{2n_0-3} \int_0^s [y^{(n_0-1)}(u)]^2 \, du,
\end{aligned}
$$

where $C = \sum_{\nu=0}^{n_0-1} a_\nu (\frac{\pi}{2})^{2(n_0-\nu-1)}$. Now we apply Lemma 8.1 with $j = n_0 - 1$, $m = n_0$, $p = \mu$, $q = 2$, $r = 2$, and $\alpha = \frac{2\mu}{\delta}$. Note that the hypotheses of this lemma are satisfied since $\mu \leq 2$. Setting $\omega = \frac{1-\alpha}{r} = \frac{\delta-2\mu}{2\delta}$, we obtain

$$
\begin{aligned}
z(s) &\leq C s^{2n_0-3} \int_0^\infty [Y_s^{(n_0-1)}(u)]^2 \, du \\
&\leq C M^2 s^{2n_0-3} \left[\int_0^\infty |Y_s(u)|^\mu \, du\right]^{\frac{4}{\delta}} \left[\int_0^\infty [Y_s^{(n_0)}(u)]^2 \, du\right]^{2\omega} \\
&\leq C M^2 s^{2n_0-3} \left[\int_0^s |y(u)|^\mu \, du + 1\right]^{\frac{4}{\delta}} \left[\int_0^s [y^{(n_0)}(u)]^2 \, du + 1\right]^{2\omega}.
\end{aligned}
$$

Lemma 8.4 and (8.8) yield

$$
\begin{aligned}
z(s) &\leq C M^2 s^{2n_0-3}(K + 1)^{\frac{4}{\delta}} \left[\int_0^s z^{(n)}(u) \, du + 1\right]^{2\omega} \\
&\leq K_1 s^{2n_0-3}[z^{(n-1)}(s) - z^{(n-1)}(0) + 1]^{2\omega} \leq K_1 s^{2n_0-3}(z^{(n-1)}(s))^{2\omega},
\end{aligned}
$$

where $K_1 = C M^2 (K + 1)^{\frac{4}{\delta}}$. Thus,

$$z^{(n-1)}(s) \geq \left(\frac{z(s)}{K_1}\right)^{\frac{1}{2\omega}} s^{-\frac{(n-3)}{2\omega}}, \quad s > 0. \tag{8.12}$$

Next, we show that the inequality (8.12) has no positive nondecreasing solutions on $(0, \infty)$. From (8.6) and the definition of $z$, it follows that

$$z(t) = z(t) - z(0) = \int_0^t z'(s) \, ds \leq z'(t) t, \quad t > 0,$$

so

$$z'(t) \geq \frac{z(t)}{t}, \quad t > 0.$$

Similarly,

$$z^{(j)}(t) \geq \frac{z(t)}{t^j}, \quad t > 0, \quad j = 1, \ldots, n - 2. \tag{8.13}$$

In view of (8.11), we can apply Lemma 8.2 to $z$ with $m = n - 1$ and $\alpha = \frac{\delta - 2\mu}{\delta(n-1) - \mu n}$. Then, by Lemma 8.2, (8.12), and (8.13), we obtain

$$\left(\frac{t^{\frac{n-2}{2}}}{z(t)}\right)^{\sigma} \geq \prod_{i=0}^{n-2} |z^{(i)}(t)|^{-\varepsilon} \geq \gamma \int_t^{\infty} [z^{(n-1)}(s)]^{\alpha} [z(s)]^{\beta} \, ds$$

$$\geq \gamma K_2 \int_t^{\infty} z^k(s) s^{-l} \, ds \geq \gamma K_2 \int_t^{\infty} \frac{ds}{K s^l} = \infty$$

for $t > 0$, where $\sigma = \frac{2(1 - n\alpha + \alpha)}{n-2} > 0$, $\beta = \frac{n\alpha - 2}{n-2}$, $K_2 = K_1^{-\frac{\alpha}{2\omega}}$, $k = \frac{\alpha}{2\omega} + \beta = 0$, and $l = \frac{(n-3)\alpha}{2\omega} \leq 1$. Thus, $z \equiv 0$ on $\mathbb{R}_+$, which contradicts (8.11), and so (8.8) cannot hold. $\qquad\square$

**Corollary 8.1.** *Let* $n = 4k$. *If* (8.2)–(8.5) *hold, then equation* (8.1) *is of the nonlinear limit–point type.*

**Theorem 8.3.** *Assume that* $n \geq 4$, (8.3) *holds, there exists* $M > 0$ *such that*

$$-M \leq r(t) \leq 0,$$

*and there exist positive constants* $M_i$, $i = 1, 2$, *and* $\lambda \in (0, 1]$ *such that*

$$M_1 |x_1|^{\lambda} \leq |f(x_1, x_2, \ldots, x_n)| \leq M_2 |x_1|^{\lambda} \quad on \quad \mathbb{R}^n.$$

*Then equation* (8.1) *is of the nonlinear limit–point type.*

*Proof.* Let $y$ be a solution of (8.1) having the initial conditions (8.6). If $y$ is nonoscillatory, the proof is similar to the proof of Theorem 8.2. So assume that $y$ is an oscillatory solution defined on $\mathbb{R}_+$ and let $\mu = 1 + \lambda$. First, we prove that

$$\int_0^{\infty} |y(t)|^{\mu} \, dt = \infty. \tag{8.14}$$

Suppose

$$\int_0^{\infty} |y(t)|^{\mu} \, dt = C < \infty. \tag{8.15}$$

Let $\sigma = \frac{1+\lambda}{\lambda}$; then, we have

$$\int_0^\infty [y^{(n)}(s)]^\sigma \, ds = \lim_{t \to \infty} \int_0^t r^\sigma(s) \Big( f(y(s), \ldots, y^{(n-1)}(s)) \Big)^\sigma \, ds$$

$$\leq M^\sigma M_2^\sigma \lim_{t \to \infty} \int_0^t |y(s)|^\mu \leq C_1 < \infty, \tag{8.16}$$

where $C_1$ is a suitable constant. Thus, (8.15), (8.16), and Lemma 8.1 with $j = 1, 2$, $m = n$, $p = \mu$, $q = 3(1 + \lambda) = 3\mu$, and $r = \sigma$, yield

$$y^{(j)} \in L^{3\mu}(\mathbb{R}), \qquad j = 1, 2. \tag{8.17}$$

Define

$$Y(t) = \begin{cases} y(t), & \text{for } t \in [0, \infty), \\ 0, & \text{for } t \in (-\infty, 0). \end{cases}$$

According to (8.6), $Y'$ and $Y''$ are continuous, and $Y'$, $Y'' \in L^{3\mu}(\mathbb{R})$; moreover, Lemma 8.7 implies that $Y'$, and hence $y'$, is bounded. This contradiction to Lemma 8.6 proves that (8.14) holds. Finally,

$$\int_0^\infty y(t) \, f(y(t), \ldots, y^{(n-1)}(t)) \, dt \geq M_1 \int_0^\infty |y(t)|^\mu \, dt = \infty,$$

and so $y$ is of the nonlinear limit–point type. $\qquad\qquad\qquad\qquad\qquad\square$

**Remark 8.2.** The conclusion of Theorem 8.3 was proved for third order equations under a weaker nonlinearity condition on $f$ (see Theorem 6.3 above).

The following two theorems generalize the nonlinearity condition imposed on $f$ in Theorem 8.3.

**Theorem 8.4.** *Let $n = 2k + 1$ and suppose that there exist constants $M_1 > 0$ and $M_2 > 0$, such that*

$$M_1 \leq -r(t) \leq M_2,$$

*and there is a positive constant $M$ and a continuous function $g : \mathbb{R}_+ \to \mathbb{R}$ with $g(0) = 0$, $g(x) > 0$ for $x > 0$, $\liminf_{x \to \infty} g(x) > 0$, and*

$$g(|x_1|) \leq |f(x_1, \ldots, x_n)| \leq M(1 + |x_1|) \qquad \text{on} \quad \mathbb{R}^n.$$

*Then equation (8.1) is of the nonlinear limit–point type.*

*Proof.* Consider a solution of (8.1) having the initial conditions

$$y^{(i)}(0) > 0, \quad i = 0, 1, \ldots, n-1, \quad z^{(n-1)}(0) > 0,$$

where the function $z$ is defined in Lemma 8.4. The conclusion can be proved in a manner similar to the proof of Theorem 8.2 if $y$ is nonoscillatory. Now suppose that $y$ is an oscillatory solution. It follows from Lemma 8.4 that $\lim_{t \to \infty} z^{(n-1)}(t) = \infty$ and

$$\infty = -\int_0^\infty y(t)\, y^{(n)}(t)\, dt \le \int_0^\infty |r(t)| y(t) f(y(t), \ldots, y^{(n-1)}(t))\, dt$$

$$\le M_2 \int_0^\infty y(t)\, f(y(t), \ldots, y^{(n-1)}(t))\, dt,$$

and so $y$ is of the nonlinear limit–point type. $\qquad\square$

**Theorem 8.5.** *Let* $n = 4k$, *suppose* (8.3) *holds,* $r \le 0$ *on* $\mathbb{R}_+$, *and there exist constants* $K_i$, $i = 0, 1, 2, 3, 4$, *and* $x^*$ *such that*

$$|r(t)| \le K_0 t^\delta$$

*and*

$$g_1(|x_1|) \le |f(x_1, \ldots, x_n)| \le g_2(|x_1|) \quad \text{on} \quad \mathbb{R}^n, \tag{8.18}$$

*where* $\delta = \frac{n+1}{n-2}$,

$$g_1(x) = \begin{cases} K_1 x, & \text{for } x \in [0, x^*], \\ K_2, & \text{for } x \in (x^*, \infty), \end{cases}$$

*and*

$$g_2(x) = \begin{cases} K_3, & \text{for } x \in [0, x^*], \\ K_4 x, & \text{for } x \in (x^*, \infty). \end{cases}$$

*Then equation* (8.1) *is of the nonlinear limit–point type.*

*Proof.* Consider a solution $y$ of (8.1) with initial conditions (8.6). Then (8.11) holds and the proof is similar to the proof of Theorem 8.2 if $y$ is nonoscillatory. Hence, we assume that $y$ is oscillatory.

*Case (i).* Suppose that $y$ is unbounded. According to [5, Lemmas 1 and 2], there exist sequences $\{\alpha_k\}_1^\infty$ and $\{\beta_l\}_0^\infty$ with $\beta_0 = 0$ such that

$$\alpha_k < \beta_k, \quad |y(\alpha_k)| = |y(\beta_k)| = x^*, \quad |y(t)| > x^* \text{ for } t \in (\alpha_k, \beta_k),$$

and

$$|y(t)| \le x^* \text{ for } t \in [\beta_{k-1}, \alpha_k], \quad k = 1, 2, \ldots.$$

For $t \geq 0$, define

$$y_1(t) = \begin{cases} y(t), & \text{for } |y(t)| \leq x^*, \\ 0, & \text{otherwise,} \end{cases}$$

$$y_2(t) = \begin{cases} 0, & \text{for } |y(t)| \leq x^*, \\ y(t), & \text{otherwise,} \end{cases}$$

and

$$Y_1(t) = \begin{cases} y_1(t), & \text{for } |y(t)| \leq x^*, \\ v(t), & \text{otherwise,} \end{cases} \quad \begin{pmatrix} \text{i.e., } t \in \bigcup_{k=1,2,\dots}^{\infty} [\beta_{k-1}, \alpha_k] \end{pmatrix} \\ \begin{pmatrix} \text{i.e., } t \in \bigcup_{k=1,2,\dots}^{\infty} (\alpha_k, \beta_k) \end{pmatrix}$$

$$Y_2(t) = \begin{cases} y_2(t), & \text{for } |y(t)| \geq x^*, \\ w(t), & \text{otherwise,} \end{cases}$$

where $v, w$ are such that $Y_i \in \tilde{C}^{n-1}(\mathbb{R}_+)$, $i = 1, 2$, and

$$\sum_{k=1}^{\infty} \int_{\alpha_k}^{\beta_k} v^2(s)\,ds \leq 1, \quad \sum_{k=1}^{\infty} \int_{\alpha_k}^{\beta_k} |v^{(n)}(s)|\,ds \leq 1,$$

$$\sum_{k=1}^{\infty} \int_{\beta_{k-1}}^{\alpha_k} |w(s)|\,ds \leq 1, \quad \sum_{k=1}^{\infty} \int_{\beta_{k-1}}^{\alpha_k} |w^{(n)}(s)|\,ds \leq 1.$$

Note that

$$\int_0^t Y_1^2(s)\,ds \leq \int_0^t y_1^2(s)\,ds + 1, \quad \int_0^t |Y_2(s)|\,ds \leq \int_0^t |y_2(s)|\,ds + 1,$$

and

$$\int_0^t |Y_i^{(n)}(s)|\,ds \leq \int_0^t |y_i^{(n)}(s)|\,ds + 1, \quad t \in [0, \infty), \quad i = 1, 2.$$

For $k \in \mathbb{N} = \{1, 2, \dots\}$ and $i = 1, 2$, define

$$Y_{ik}(t) = \begin{cases} Y_i(t) & \text{for } t \in [0, \beta_k], \\ u_i(t) & \text{for } t \in (\beta_k, \infty), \end{cases}$$

where $u_i$, $i = 1, 2$ are suitably chosen functions such that $Y_{ik} \in \tilde{C}^{n-1}(\mathbb{R}_+)$,

$$\int_{\beta_k}^{\infty} |u_i^{(n)}(s)|\,ds \leq 1, \quad \int_{\beta_k}^{\infty} u_1^2(s)\,ds \leq 1, \quad \text{and} \quad \int_{\beta_k}^{\infty} |u_2(s)|\,ds \leq 1.$$

Now (8.11) and Lemma 8.4 imply $\lim_{t \to \infty} z^{(n-1)}(t) = \infty$, and thus for $M_1 > 0$ there exists $t_1 \geq 0$ such that

$$M_1 t^{n-1} \leq z(t), \quad t \in [t_1, \infty). \tag{8.19}$$

It follows from Lemma 8.3, Lemma 8.4, and (8.6) that

$$z(t) \leq \sum_{k=1}^{n_0-1} a_k J_{2k}\left(t; [y^{(k)}]^2\right) \leq C_1 t^{n-3} \int_0^t [y^{(n_0-1)}(s)]^2 \, ds$$

$$\leq C_1 t^{n-3} \left\{ \int_0^t [Y_1^{(n_0-1)}(s)]^2 \, ds + \int_0^t [Y_2^{(n_0-1)}(s)]^2 \, ds \right\},$$

for $t \in \mathbb{R}_+$, where $C_1 = \sum_{k=1}^{n_0-1} a_k (\frac{\pi}{2})^{2(n_0-k-1)}$. Applying Lemma 8.1, we obtain

$$z(\beta_k) \leq C_1 \beta_k^{n-3} \left\{ \int_0^\infty [Y_{1k}^{(n_0-1)}(s)]^2 \, ds + \int_0^t [Y_{2k}^{(n_0-1)}(s)]^2 \, ds \right\}$$

$$\leq M^2 C_1 \beta_k^{n-3} \left\{ \left( \int_0^\infty Y_{1k}^2(s) \, ds \right)^{\gamma_1} \left( \int_0^\infty |Y_{1k}^{(n)}(s)| \, ds \right)^{\gamma_2} \right.$$

$$\left. + \left( \int_0^\infty |Y_{2k}(s)| \, ds \right)^{\gamma_3} \left( \int_0^\infty |Y_{2k}^{(n)}(s)| \, ds \right)^{\gamma_4} \right\}, \tag{8.20}$$

for $t \in \mathbb{R}_+$, where $\gamma_1 = \dfrac{n_0 + 1/2}{n - 1/2}$, $\gamma_2 = \dfrac{n-2}{n-1/2}$, $\gamma_3 = \dfrac{n+1}{n}$, and $\gamma_4 = \dfrac{n-1}{n}$.
Furthermore, there exist $k_0 \in \mathbb{N}$ and $C_2 > 0$ such that for $k \geq k_0$, we have

$$\int_0^\infty |Y_{1k}^{(n)}(s)| \, ds \leq \int_0^{\beta_k} |Y_1^{(n)}(s)| \, ds + 1 \leq \int_0^{\beta_k} |y_1^{(n)}(s)| \, ds + 2$$

$$\leq \int_0^{\beta_k} |r(s)| \, g_2(|y_1(s)|) \, ds + 2 \leq K_0 K_3 \int_0^{\beta_k} s^\delta \, ds \leq K_0 K_3 \beta_k^{\delta+1},$$

$$\int_0^\infty |Y_{2k}^{(n)}(s)| \, ds \leq \int_0^{\beta_k} |Y_2^{(n)}(s)| \, ds + 1 \leq \int_0^{\beta_k} |y_2^{(n)}(s)| \, ds + 2$$

$$\leq \int_0^{\beta_k} |r(s)| \, g_2(|y_2(s)|) \, ds + 2 \leq K_0 K_4 \int_0^{\beta_k} s^\delta |y_2(s)| \, ds + 2$$

$$\leq K_0 K_4 \beta_k^\delta \int_0^{\beta_k} |y_2(s)| \, ds + 2 \leq 2 K_0 K_4 \beta_k^\delta \int_0^{\beta_k} |y_2(s)| \, ds,$$

$$\int_0^\infty Y_{1k}^2(s)\, ds \le \int_0^{\beta_k} y_1^2(s)\, ds + 2 \le C_2 \int_0^{\beta_k} y_1^2(s)\, ds,$$

and

$$\int_0^\infty |Y_{2k}(s)|\, ds \le \int_0^{\beta_k} |y_2(s)|\, ds + 2 \le C_2 \int_0^{\beta_k} y_2(s)\, ds.$$

From this and (8.20), we obtain

$$z(\beta_k) \le C_3\, \beta_k^{n-3}\Big[\beta_k^{(\delta+1)\gamma_2}\Big(\int_0^{\beta_k} y_1^2(s)\, ds\Big)^{\gamma_1} + \beta_k^{\delta\gamma_4}\Big(\int_0^{\beta_k} |y_2(s)|\, ds\Big)^{\gamma_4+\gamma_3}\Big]$$

$$\le C_3\beta_k^{n-1}\Big[\Big(\int_0^{\beta_k} y_1^2(s)\, ds\Big)^{\gamma_1} + \Big(\int_0^{\beta_k} |y_2(s)|\, ds\Big)^2\Big],$$

for $k \ge k_0$, where $C_3 = M^2 C_1 \max\Big(C_2^{\gamma_1}(K_0 K_3)^{\gamma_2},\, C_2^{\gamma_3}(2K_0 K_4)^{\gamma_4}\Big)$.
In view of (8.19), either

$$\int_0^\infty y_1^2(s)\, ds = \infty \text{ or } \int_0^\infty |y_2(s)|ds = \infty,$$

so

$$\int_0^\infty y(t)\, f\big(y(t), \ldots, y^{(n-1)}(t)\big)\, dt = \sum_{i=1}^2 \int_0^\infty y_i(t)\, f\big(y_i(t), \ldots, y_i^{(n-1)}(t)\big)\, dt$$

$$\ge K_1 \int_0^\infty y_1^2(t)\, dt + K_2 \int_0^\infty |y_2(t)|\, dt = \infty.$$

*Case (ii).* Suppose that $y$ is bounded, i.e., $|y(t)| \le C_4$ for some $C_4 > 0$ and all $t \in \mathbb{R}_+$. Then define $f_1 : \mathbb{R}^n \to \mathbb{R}$ by

$$f_1(x_1, x_2, \ldots, x_n) = \begin{cases} f(x_1, \ldots, x_n), & \text{for } |x_1| \le C_4, \\ \frac{f(C_4, x_2, \ldots, x_n)}{C_4} x_1, & \text{for } x_1 > C_4, \\ \frac{f(-C_4, x_2, \ldots, x_n)}{C_4} |x_1|, & \text{for } x_1 < -C_4. \end{cases}$$

Then $f_1 \in C^0(\mathbb{R}^n)$, and it follows from (8.18) that

$$C_5|x_1| \le |f_1(x_1, \ldots, x_n)| \le C_6(1 + |x_1|),$$

where $C_5 = \min(K_1, \frac{K_2}{C_4})$ and $C_6 = \max(K_3, K_4, \frac{K_3}{C_4})$. At the same time, $y$ is a solution of

$$y^{(n)} = r(t)\, f_1(y, \ldots, y^{(n-1)}). \tag{8.21}$$

The hypotheses of Theorem 8.2 are satisfied for equation (8.21), and by the proof of that theorem, the solution $y$ of (8.21) with initial conditions (8.6) satisfies

$$\infty = \int_0^\infty y(t)\, f_1\big(y(t), \ldots, y^{(n-1)}(t)\big)\, dt = \int_0^\infty y(t)\, f\big(y(t), \ldots, y^{(n-1)}(t)\big)\, dt,$$

so the conclusion of the theorem follows.                                      □

**Example 8.1.** As an example of Theorems 8.4 and 8.5, consider the equation

$$y^{(n)} = r(t) \arctan y. \tag{8.22}$$

Let one of the following conditions hold:

(i) $r \geq 0$;

(ii) $n = 2k + 1$ and $M_1 \leq -r \leq M_2$, where $M_1, M_2 > 0$ are positive constants;

(iii) $n = 4k$ and $-Kt^\delta \leq r(t) \leq 0$, where $\delta = (n+1)/(n-2)$.

Then equation (8.22) has a solution satisfying

$$\int_0^\infty y(s) \arctan y(s)\, ds = \infty.$$

While Theorem 8.4 does relax the conditions on $f$ in Theorem 8.3, it only applies to equations of odd order. Similarly, Theorem 8.5 only applies to equations of order $4k$.

Next, we examine the higher order sublinear Thomas–Fermi equation

$$y^{(n)} = r(t)|y|^\lambda \operatorname{sgn} y, \quad \lambda \in (0, 1], \tag{8.23}$$

and show that under certain conditions this equation is of the nonlinear limit-point type. From Theorems 8.1–8.3, we have the following result.

**Corollary 8.2.** *(a) Let $n = 4k$. If the function $r$ satisfies (8.2), then (8.23) always has a solution $y \notin L^{1+\lambda}[0, \infty)$.*

*(b) Let $n = 2k + 3$ or $n = 4k + 2$. If $r \geq 0$ or $-M \leq r(t) \leq 0$, $M > 0$, then (8.23) always has a solution $y \notin L^{1+\lambda}[0, \infty)$.*

We close this section with a discussion of the relationship between the limit–circle property and some other asymptotic properties of solutions of the $n$-th order nonlinear equation

$$y^{(n)} + r(t) f(y) = 0, \tag{8.24}$$

where $r \in L_{\text{loc}}(\mathbb{R}_+)$, $f : \mathbb{R} \to \mathbb{R}$ is continuous, and $uf(u) > 0$ for $u \neq 0$. Our proof makes use of a very well-known result of Kiguradze (see, for example, [78], [79], or [80, Lemma 1.1]) which we state here for convenience.

**Lemma 8.8.** *Let y be an n-times continuously differentiable function defined on* $\mathbb{R}_+$ *and let* $y^{(n)}(t)$ *be of constant sign and not identically zero for all large t. Then there exist* $T > 0$ *and an integer* $\ell \in \{0, 1, 2, \ldots, n\}$ *such that for* $t \geq T$, *we have* $(-1)^{n-\ell} y(t) y^{(n)}(t) \geq 0$,

$$\ell > 0 \quad implies \quad y(t) y^{(j)}(t) \geq 0 \quad for \quad j = 0, 1, \ldots, \ell,$$

*and*

$$\ell \leq n - 1 \quad implies \quad (-1)^{j-\ell} y(t) y^{(j)}(t) > 0 \quad for \quad j = \ell + 1, \ldots, n - 1.$$

We assume that there exists $K > 0$ such that

$$|f(u)| \geq \frac{1}{|u|} \quad \text{for} \quad |u| \geq K. \tag{8.25}$$

**Theorem 8.6.** *Let* (8.25) *hold and let y be a nonlinear limit–circle type solution of equation* (8.24).

(a) *If* $r(t) \geq 0$ *and n is even, then y is oscillatory.*

(b) *If* $r(t) \geq 0$ *and n is odd, then y is either oscillatory or* $y^{(i)}(t) \to 0$ *as* $t \to \infty$ *for* $i = 0, 1, \ldots, n - 1$.

(c) *If* $r(t) \leq 0$ *and n is odd, then y is oscillatory.*

(d) *If* $r(t) \leq 0$ *and n is even, then y is either oscillatory or* $y^{(i)}(t) \to 0$ *as* $t \to \infty$ *for* $i = 0, 1, \ldots, n - 1$.

*Proof.* If $y$ is a nonlinear limit–circle solution of equation (8.24), then by Definition 2.1, $y$ is continuable. Moreover, any continuable nonoscillatory solution $y$ of (8.24) satisfies $y \neq 0$ for large t. Hence, by Lemma 8.8, any continuable nonoscillatory solution $y$ of equation (8.24) is one of the following two types:

(i)    $(-1)^i y(t) y^{(i)}(t) > 0$ for $i = 0, 1, \ldots, n - 1$ and large $t$ (a so-called Kneser solution);

(ii)   $y(t) y'(t) \geq 0$ for large $t$.

Thus, any nonoscillatory solution $y(t)$ has a limit, say $\lim_{t \to \infty} y(t) = c$. If this limit is finite and nonzero, then there exists $L > 0$ such that $y(t) f(y(t)) \geq L$ for large $t$, and so

$$\int^{\infty} y(t) f(y(t)) \, dt \geq L \int^{\infty} dt = \infty.$$

If $|c| = \infty$, then there exists $t_0 > 0$ such that $|y(t)| \geq K$ for $t \geq t_0$. Now using (8.25), a similar argument shows that $y(t)$ is again a nonlinear limit–point type solution. Notice that if $y^{(j)}(t) \to c_j \neq 0$ for some $j = 1, 2, \ldots, n - 1$, then successive integrations would show that $|y(t)| \to \infty$ as $t \to \infty$, which is impossible as above. Hence, (b) and (d) are proved.

To prove (a), we will show that if $r(t) \geq 0$ and $n$ is even, then there are no Kneser type solutions. Without loss of generality, assume that $y(t) > 0$ is a Kneser solution of equation (8.24). Since $n$ is even, $y^{(n-1)}(t) < 0$ and $y^{(n-2)}(t) > 0$ for all large $t$. From equation (8.24), we have $y^{(n)}(t) \leq 0$, so $y^{(n-1)}$ is negative and nonincreasing. Thus, there exists $K_1 > 0$ such that $y^{(n-1)}(t) \leq -K_1$. An integration shows that $y^{(n-2)}(t) < 0$ for large $t$, which is a contradiction.

A similar argument shows that there are no Kneser solutions if $r(t) \leq 0$ and $n$ is odd.                                                                                    □

**Remark 8.3.** Notice that condition (8.25) holds, for example, if $f(u)$ is bounded away from zero whenever $u$ is bounded away from zero.


**Open Problems.**

**Problem 8.1.** *In Theorem 6.2 and Remark 6.1 it was shown that for $n = 3$, if $r(t) \geq K > 0$ and there exists $\beta > \frac{3}{2}$ such that*

$$|f(x_1, x_2, x_3)| \leq \frac{1}{|x_1|^\beta} \quad \text{for } |x_1| \geq M > 0,$$

*then (8.1) is of the nonlinear limit–circle type. Whether this result is true for $n > 3$ remains an open question.*

**Problem 8.2.** *Extend the results in Theorem 8.6 to equations with quasiderivatives.*

**Problem 8.3.** *Other than for Theorem 8.6, the relationship between the nonlinear limit–circle property and the boundedness, oscillation, and convergence to zero of solutions for higher order equations, i.e., equations of order greater than two, has not been studied. This would appear to be a very interesting area for research.*

**Problem 8.4.** *Grimmer and Patula [69] show that if (5.1) holds, equation (1.7) is limit–circle, and $e \in L^2$, then the equation*

$$(a(t)y')' + r(t)y = e(t)$$

*has at most one nonoscillatory solution. Extend this result to higher order equations.*

**Problem 8.5.** *A number of authors have considered the problem of perturbing a limit–circle type linear equation with a nonlinear perturbation term. For example, Wong [120] compared an unperturbed n-th order linear equation to the same equation with a perturbation term of the form (4.26) above. Extensions of these types of results would be of interest.*

**Notes.** The results in this chapter are based on the paper [14] of Bartušek, Došlá, and Graef. Theorem 8.6 is new.

# Chapter 9

# Relationship to Spectral Theory

Here, we relate the study of the limit–point/limit–circle problem to the spectral theory of self-adjoint linear equations.

## 9.1. Introduction

In Section 1.2, we discussed how the limit–point/limit–circle problem for second order linear equations leads to the study of the higher order self-adjoint equation

$$\ell(y) \equiv \sum_{i=0}^{n} (-1)^i \left( p_i(t) y^{(i)} \right)^{(i)} = \lambda y \quad (\operatorname{Im} \lambda \neq 0), \tag{9.1}$$

for $t \in \mathbb{R}_+$. We also noted that the deficiency index $m$ of the minimal operator $L_0$ associated to $\ell$ is the number of linearly independent solutions of equation (9.1) that belong to $L^2$, and $n \leq m \leq 2n$. If $m = 2n$, equation (9.1) is said to be of the *limit–circle type*. If $m = n$, some authors refer to this as the *limit–point case* for (9.1). We refer the reader to the monographs [37, 57, 94, 113] for further details concerning the spectral theory of ordinary differential operators.

In this chapter, we apply our results for nonlinear equations obtained in Chapters 7 and 8 to self-adjoint linear equations, and in some cases obtain what appears to be new results for linear equations.

Kauffman, Read, and Zettl [77, p. 45] noted that "there are no known examples of functions $r$ such that

$$y^{(2k)} + r(t) y = 0 \tag{9.2}$$

is limit–circle for $k > 1$." Thus, we begin with the following conjecture.

**Conjecture 9.1.** *The equation* (9.2) *always has a solution* $y \notin L^2(\mathbb{R}_+)$, *i.e.,* (9.2) *is never of the limit–circle type.*

We will show that our conjecture is true if $r \leq 0$ or if $r$ is a bounded function (we already did this for $k = 2$ in Corollary 7.1). The first case follows from the basic theory of linear differential equations, and the second from spectral theory (see, for example, [94]). In this chapter, we will also prove that this conjecture is true for equations with $k = 2j$, $j = 1, 2, \ldots$, as long as $r$ is not an unbounded oscillatory function (see Theorem 9.2 below).

The self-adjoint linear differential equation

$$\sum_{i=1}^{n} (-1)^i \left( p_i(t) y^{(i)} \right)^{(i)} + r(t)y = 0, \tag{9.3}$$

where $p_i$, $i = 1, 2, \ldots, n$, and $r$ are continuous on $\mathbb{R}_+$, plays an important role in the spectral theory of singular differential operators; see, for example, [33, 37, 38, 39, 41, 46, 53, 94].

**Lemma 9.1.** ([94, §23, Theorem 1, p.192]) *Let $q$ be a real, measurable, essentially bounded function on $\mathbb{R}_+$. Then, the deficiency index of the expression $\ell$ is not changed by adding the function $q$ to $p_0$.*

**Theorem 9.1.** *If $r$ is a continuous and bounded function, then equation (9.3) is not limit–circle.*

*Proof.* The equation

$$\sum_{i=1}^{n} (-1)^i \left( p_i(t) y^{(i)} \right)^{(i)} = 0$$

is never of the limit–circle type since $y(t) \equiv 1 \notin L^2$ is a solution. The conclusion then follows from Lemma 9.1 .                                                                $\square$

## 9.2. Self-Adjoint Linear Fourth Order Equations

Many results in the literature are devoted to the special case of (9.1) where $n = 2$ and the coefficients are powers of $t$, say

$$y^{(4)} - (at^\alpha y')' + bt^\beta y = \lambda y \quad (\text{Im } \lambda \neq 0), \tag{9.4}$$

where $a, b, \alpha, \beta \in \mathbb{R}$. As can be seen from the discussion by Walker [111], it is known that (9.4) is not of the limit–circle type for any values of $\alpha$ and $\beta$ except perhaps where (i) $\beta \leq 0$ and $0 < \alpha < 2$ or (ii) $\beta > 0$ and $\beta/2 < \alpha < \beta + 2$. Theorem 9.1 above takes care of (i).

As a consequence of the Theorem in [112], Walker (see [112, p. 332]), concluded that equation (9.4) is limit–circle if

$$a < 0, b > 0, \alpha + \beta > 2, \text{ and } \alpha - 2 < \beta < 2\alpha.$$

(In this regard, also see the discussion in Eastham [40, p. 55].)

**Definition 9.1.** The linear equation (9.4) is said to be *limit–m* if it has exactly $m$ linearly independent solutions belonging to $L^2$.

Devinatz [33] (in particular, see Figures 1 and 2 on p. 283) gives a very careful analysis of what is known for equation (9.4) in case $ab > 0$ or $ab < 0$. In particular, he shows that if $ab > 0$, then (9.4) is known to be either *limit–2* or *limit–3* except along the line $\beta = \alpha - 2$; along this line, the deficiency index is not known. (In this regard, also see the discussion in Eastham and Grudniewicz [41].) If $ab < 0$, Devinatz shows that (9.4) is either *limit–2* or *limit–3* exterior to the region

$$\begin{cases} \beta \geq 0, \\ \beta/2 \leq \alpha \leq \beta + 2, \\ \alpha + \beta > 2. \end{cases} \tag{9.5}$$

The deficiency index is not known along the boundary line $\beta = \alpha - 2$ (again see Eastham and Grudniewicz [41]) and depends on the values of $a$ and $b$ along the boundary line $\beta = 2\alpha$. Moreover, if $a > 0$, then (9.4) is *imit–3* in the interior of this region, and if $a < 0$, then (9.4) is *limit–4*, i.e., limit–circle, in the interior of the region. This is in complete agreement with Walker [112, p. 332].

In order to obtain additional information, let us apply Theorem 7.7 above to the boundary line $\beta = \alpha - 2$ with $\beta > 0$. On this line we have that $z(t) = b/(b - a\alpha)$. It is easy to see that $a > 0$ and $a\alpha > b$ are needed to satisfy the hypotheses of Theorem 7.7. That is, on the line $\beta = \alpha - 2$ with $\beta > 0$, equation (9.4) will not be of the limit–circle type if the added condition

$$a > 0 \text{ and } a\alpha > b$$

holds. When $b > 0$ so that $ab > 0$, this is not much of a surprise in view of Devinatz's results [33] that this is a boundary line between a *limit–2* and a *limit–3* region. If $b < 0$ so that $ab < 0$, this result appears to be new. The analysis needed to apply Theorem 7.7 to the other boundary line $\beta = 2\alpha$ is more difficult and is left to the reader. (An alternate approach is given in the proof of part (vi) of Corollary 9.1 below.) The above remarks are summarized in the following corollary.

**Corollary 9.1.** *Equation* (9.4) *is not limit–circle if any of the following conditions hold:*

 (i) *ab* > 0 *and either*

   (a) $\beta \leq 0$, *or*

   (b) $\beta > 0$ *and* $\beta \neq \alpha - 2$;

 (ii) *ab* > 0, $\beta > 0$, $\beta = \alpha - 2$, *and* $a\alpha > b$;

 (iii) *ab* = 0;

 (iv) *ab* < 0, *a* > 0, $\alpha + \beta > 2$, $\beta \neq 2\alpha$, *and* $\beta \neq \alpha - 2$;

 (v) *ab* < 0, *a* > 0, $\beta \geq 0$, $\beta = \alpha - 2$ *and* $a\alpha > b$;

 (vi) $\beta = 2\alpha$ *and either*

   (a) $0 > a \geq -2\sqrt{|b|}$, *or*

   (b) $a < -2\sqrt{|b|}$ *and* $\alpha \leq \frac{2}{3}$.

*Proof.* Part (i) follows from Devinatz's paper [33, §4]; part (ii) follows from [33, §4] and our discussion of the boundary line $\beta = \alpha - 2$ given above. If *b* = 0, part (iii) follows immediately; if *a* = 0, then this result follows from Theorem 9.1 above. Part (iv) follows from [33, §4] and (v) from our discussion of the boundary line $\beta = \alpha - 2$.

Part (vi) follows from [53, Theorem 5.1] where it is also required that the equation $x^4 + cx^2 + 1 = 0$ with $c = |a|/\sqrt{|b|}$ does not have four distinct purely imaginary zeros. This restriction is satisfied if $0 > a \geq -2\sqrt{|b|}$. This stipulation will also be satisfied if $a < -2\sqrt{|b|}$, but in this case [53, Theorem 5.1] also requires an additional condition on $\alpha$, namely, (1.13) in [53], i.e., $\alpha \leq \frac{2}{3}$. (Also see the discussion in [33, p. 284].)                                        □

**Remark 9.1.** From Corollary 9.1 (vi) above, it follows that if $a < 0$, $|a| > 2\sqrt{b} > 0$, and $\beta = 2\alpha > \frac{4}{3}$, then equation (9.4) is limit–circle. That is, on the boundary $\beta = 2\alpha$ of the region (9.5), whether or not equation (9.4) is limit–circle actually depends on the size of the coefficients *a* and *b* and not just on their sign.

**Remark 9.2.** Relative to parts (i) and (iv) in Corollary 9.1, Kauffman [76, Theorem 2.5] shows that equation (9.4) is not limit–circle if $a \geq 0$, $b \geq 0$, and $\beta > \alpha - 2$. On the other hand, with $a \geq 0$, $b \leq 0$, and no restrictions on the sign of $\beta$, Corollary 3 (i) in [13] guarantees this same conclusion holds. (The case $a = 0$ and $b > 0$ is still of interest and is included as a special case of the discussion in Section 9.3

below.) This offers an alternate verification to parts of our corollary and suggests a possible simplification in the listing of the number of parts. However, we chose the previous approach in order to indicate some of the rich history of this problem.

**Remark 9.3.** Eastham [40] shows that equation (9.4) is of the limit–circle type if $a < 0, b > 0, \beta = \alpha - 2 > 0$, and

$$\left(\frac{b}{-a}\right)^{\frac{1}{2}} > \frac{(3\beta + 2)(\beta + 2)}{8\beta}.$$

The following equation is an example of equation (9.4) that is of the limit–circle type.

**Example 9.1.** Consider the equation

$$y^{(4)} + (at^\alpha y')' + bt^{2\alpha} y = 0.$$

If $a > 0$ and $b > 0$, then this equation has all its solutions in $L^2$, i.e., the equation is of the limit–circle type, if and only if $a > 2\sqrt{b}$ and $\alpha > \frac{2}{3}$. For the proof, see [53, Theorem 5.1] and Corollary 9.1.

Now, we apply results from Section 7.2 to the self-adjoint linear differential equation

$$Ly \equiv (p_2(t)y'')'' - (p_1(t)y')' + p_0(t)y = \lambda y \quad (\text{Im } \lambda \neq 0), \qquad (9.6)$$

where $p_i \in C^i[a, \infty), i = 0, 1, 2$, are real-valued functions and $p_2 > 0$.
The following conjecture is still open (see, for example, [97, 102]).

**Conjecture 9.2.** *Real formally self-adjoint expressions with nonnegative coefficients are not limit–circle.*

In our case, this means that the deficiency index of $L$ with $p_i \geq 0, i = 0, 1$, is either 2 or 3. This conjecture was proved by Kauffman [76] provided that the coefficients are finite sums of real multiples of real powers satisfying certain other conditions. Theorem 9.1 proves this conjecture for the fourth order equation (9.6) with bounded $p_0$ and arbitrary $p_1$. In case $p_0 \leq 0$ and is unbounded, the following result holds; it follows from Theorem 7.3 above.

**Corollary 9.2.** *Equation (9.6) is limit–2 or limit–3 if $p_0 \leq 0$ and equation (7.28) is nonoscillatory.*

When $p_1 \equiv 0$ and $p_0 \geq 0$, we will revisit this problem in the next section.

## 9.3. Two-Term Even Order Linear Equations

We will apply Theorems 8.1 and 8.2 to the linear equation

$$My \equiv y^{(4k)} + r(t)y = 0. \tag{9.7}$$

In Section 9.1, it is noted that *there are no known examples of functions r such that the differential expression M is limit–circle, i.e., all solutions of (9.7) are in $L^2$.* As a consequence of our results, we can clarify this situation. As long as $r$ does not change sign or is an oscillatory function that is either bounded from above or bounded from below, (9.7) can never be a limit–circle equation. This result is formalized below; the first corollary is an immediate consequence of Theorems 8.1 and 8.2.

**Theorem 9.2.** *If the function r satisfies (8.2) or is an oscillatory function that is either bounded from above or bounded from below, then equation (9.7) is not of the limit–circle type.*

*Proof.* If $r$ satisfies (8.2), then the conclusion follows immediately from Corollary 8.1. Suppose that $r$ is an oscillatory function that is bounded from above (below). Then there exists a constant $k < 0$ $(k > 0)$ such that $r < -k$ $(r > -k)$. By Corollary 8.1,

$$y^{(4k)} + (r(t) + k)y = 0$$

is not limit–circle. By Lemma 9.1, the equation

$$y^{(4k)} + (r(t) + k + q(t))y = 0$$

is not limit–circle whenever $q$ is a measurable and essentially bounded function. Thus, letting $q = -k$ we obtain that (9.7) is also not of the limit–circle type.    □

**Remark 9.4.** Evans and Zettl [44] arrive at the same conclusion as in Theorem 9.2 by studying what are known as "strong interval type" limit–point criteria (also see Everitt [48]).

**Corollary 9.3.** *The differential equation*

$$y^{(4)} - t^{\frac{4}{3}+\delta}y = \lambda y, \quad \delta > 0 \tag{9.8}$$

*is limit–3.*

*Proof.* By a result in [46, §7], (9.8) is limit–$\nu$, where $\nu \in \{3, 4\}$. By Theorem 9.2, it follows that $\nu \in \{2, 3\}$.    □

**Remark 9.5.** Corollary 9.3 also follows from [39, Theorem 1] with $S = h = 0$, or the Corollary in [39, p. 433] with $k = 0$, where a completely different (i.e., asymptotic) method has been used.

**Example 9.2.** Consider the differential expression

$$Ry \equiv y^{(4)} + t^\alpha \sin t^\beta y$$

where either (i) $\alpha \geq 0$ and $\beta \leq 0$ or (ii) $\alpha \leq 0$ and $\beta \in \mathbb{R}$. Then, by Theorem 9.2 $R$ is not limit–circle. This result can also be seen as a special case of [77, p. 94].

We conclude this section by noting the implication that the above results have on the study of the nonlinear limit–point/limit–circle problem for equations of the form

$$y^{(4)} + r(t)f(y) = 0. \tag{9.9}$$

As a consequence of Theorem 9.2, unless $r$ is an unbounded oscillatory function, it would not be possible to find sufficient conditions for equation (9.9) to be of the nonlinear limit–circle type if the conditions on the nonlinear function $f$ include linear functions as a special case. This is not the case for second order equations as can be seen from results in Chapter 3.

In view of the above results, we propose the following problems for future research.

**Open Problems**

**Problem 9.1.** *Analyze the situation along the boundary line $\beta = 2\alpha$ in the region described by (9.5) as was done there.*

**Problem 9.2.** *Does a result analogous to Theorem 6.2 hold for equations of order $n > 3$?*

**Problem 9.3.** *Under what conditions, such as $|r(t)| \leq |R(t)|$ for all $t > t_0$, is the following statement true.*
*If*

$$y^{(4)} - (p(t)y')' + R(t)y = 0$$

*is not limit–circle, then*

$$y^{(4)} - (p(t)y')' + r(t)y = 0$$

*is not limit–circle. If this statement is not true, then construct a counterexample.*

*In view of Theorem 9.1, in order for this problem to be of interest, it should be assumed that r is an unbounded function. Moreover, if p $\equiv$ 0, then r should be assumed to be oscillatory as well (see Theorem 9.2). (This is also noted by Hallam [70, p. 138] for second order equations.) In this regard, see the Appendix in the paper by Hartman and Wintner [74] (also see the paper of Eastham and Thompson [42]) where they construct an example for second order equations to show that it is possible to have the equation*

$$y'' + R(t)y = 0$$

*being of the limit–point type with*

$$R(t) \geq r(t) \geq 0$$

*and yet the equation*

$$y'' + r(t)y = 0$$

*is of the limit–circle type.*

**Problem 9.4.** *Ladas [91] studied the relationship between oscillation and the spectrum of self-adjoint linear differential operators of order 2n. The relationship to boundedness and convergence to zero would be interesting topics for future research.*

**Problem 9.5.** *Does there exist an oscillatory function r that is unbounded from above and below such that all solutions of equation (9.7) belong to $L^2$?*

**Problem 9.6.** *Resolve Conjecture 9.2 for the case $p_1 \not\equiv 0$ with $p_0$ being an unbounded function.*

**Notes.** Theorem 9.2 is due to Bartušek, Došlá, and Graef [15].

# Bibliography

[1] F. V. Atkinson, Nonlinear extensions of limit–point criteria, *Math. Z.* **130** (1973), 297–312.

[2] M. Bartušek, *Asymptotic Properties of Oscillatory Solutions of Differential Equations of the n-th Order*, Folia FSN Univ. Brunensis Masarykianae, (1992).

[3] M. Bartušek, On the structure of solutions of a system of three differential inequalities, *Arch. Math. (Brno)* **30** (1994), 117–130.

[4] M. Bartušek, On structure of solutions of a system of four differential inequalities, *Georg. Math. J.* **2** (1995), 225–236.

[5] M. Bartušek, Oscillatory criteria for nonlinear $n$-th order differential equations with quasiderivatives, *Georg. Math. J.* **3** (1996), 301–314.

[6] M. Bartušek, Asymptotic behavior of oscillatory solutions of $n$-th order differential equations with quasiderivatives, *Czech. Math. J.* **47** (**122**) (1997), 245–259.

[7] M. Bartušek, M. Cecchi, Z. Došlá, and M. Marini, On nonoscillatory solutions of third order nonlinear differential equations, *Dynamic Systems Appl.* **9** (2000), 483–500.

[8] M. Bartušek and Z. Došlá, On solutions of a third order nonlinear differential equation, *Nonlinear Anal.* **23** (1994), 1331–1343.

[9] M. Bartušek and Z. Došlá, Oscillatory criteria for nonlinear third order differential equations with quasiderivatives, *Differential Eqs. Dynam. Systems* **3** (1995), 251–268.

[10] M. Bartušek and Z. Došlá, Remark on Kneser problem, *Appl. Anal.* **56** (1995), 327–333.

[11] M. Bartušek and Z. Došlá, On the limit–point/limit–circle problem for nonlinear third order differential equations, *Math. Nachr.* **187** (1997), 5–18.

[12] M. Bartušek, Z. Došlá, and J. R. Graef, On $L^2$ and limit–point type solutions of fourth order differential equations, *Appl. Anal.* **60** (1996), 175–187.

[13] M. Bartušek, Z. Došlá, and J. R. Graef, Limit–point type results for nonlinear fourth order differential equations, *Nonlinear Anal.* **28** (1997), 779–792.

[14] M. Bartušek, Z. Došlá, and J. R. Graef, Nonlinear limit–point type solutions of $n$-th order differential equations, *J. Math. Anal. Appl.* **209** (1997), 122–139.

[15] M. Bartušek, Z. Došlá, and J. R. Graef, The nonlinear limit–point/limit–circle problem for higher order equations, *Arch. Math. (Brno)* **34** (1998), 13–22.

[16] M. Bartušek, Z. Došlá, and J. R. Graef, On the definitions of the nonlinear limit–point/limit–circle properties, *Differential Eqs. Dynamical Syst.* **9** (2001), 49–61.

[17] M. Bartušek and J. R. Graef, On $L^2$ solutions of third order nonlinear differential equations, *Dynam. Systems Appl.* **9** (2000), 469–482.

[18] M. Bartušek and J. R. Graef, Some limit–point/limit–circle results for third order differential equations, *Disc. Cont. Dyn. Syst.*, Proceedings of the Third International Conference on Dynamical Systems and Differential Equations, 31–38 (2001).

[19] M. Bartušek and J. R. Graef, On the limit-point/limit-circle problem for second order nonlinear equations, *Nonlinear Stud.* **9** (2002), 361–369.

[20] E. F. Beckenbach and R. Bellman, *Inequalities*, Springer-Verlag, Berlin (1961).

[21] R. Bellman, *Stability Theory of Differential Equations*, McGraw-Hill, New York (1953).

[22] O. Borůvka, *Linear Differentialtransformationen 2. Ordung*, VEB Deutscher Verlag, Berlin (1967).

[23] J. Burlak, On the non–existence of $L_2$-solutions of nonlinear differential equations, *Proc. Edinburgh Math. Soc.* **14** (1965), 257–268.

[24] T. A. Burton and W. T. Patula, Limit circle results for second order equations, *Monatsh. Math.* **81** (1976), 185–194.

[25] M. Cecchi, Z. Došlá, and M. Marini, On third order differential equations with property A and B, *J. Math. Anal. Appl.* **231** (1999), 509–525.

[26] M. Cecchi, M. Marini, and G. Villari, Integral criteria for a classification of solutions of linear differential equations, *J. Differential Equations* **99** (1992), 381–397.

[27] T. A. Chanturia, On Kneser problem for systems of ordinary differential equations, *Mat. Zametki* **15** (1974), 897–906.

[28] T. A. Chanturia, On singular solutions of nonlinear systems of ordinary differential equations, (Colloq., Keszthely, 1974), *Colloq. Math. Soc. János Bolyai* **15** (1976), 107–119.

[29] T. A. Chanturia, On existence of singular and unbounded oscillatory solutions of differential equations of the Emden–Fowler's type, *Diff. Urav.* **28** (1992), 1009–1022. (In Russian).

[30] E. A. Coddington and N. Levinson, *Theory of Ordinary Differential Equations*, McGraw–Hill, New York, (1955).

[31] W. A. Coppel, *Stability and Asymptotic Behavior of Differential Equations*, Heath, Boston (1965).

[32] J. Detki, The solvability of a certain second order nonlinear ordinary differential equation in $L^p(0, \infty)$, *Math. Balk.* **4** (1974), 115–119. (In Russian).

[33] A. Devinatz, The deficiency index of certain fourth–order ordinary self adjoint differential operators, *Quart. J. Math. Oxford (2)* **23** (1972), 267–286.

[34] A. Devinatz, The deficiency index problem for ordinary self adjoint differential operators, *Bull. Amer. Math. Soc.* **79** (1973), 1109–1127.

[35] Z. Došlá, On oscillatory solutions of third-order linear differential equations, *Čas. Pěst. Mat.* **114** (1989), 28–34.

[36] Z. Došlá, On square integrable solutions of third order linear differential equations, *Inter. Scientific Conf. Math. (Herl'any 1999), Univ. Technology Košice*, (2000), 68–72.

[37]  N. Dunford and J. T. Schwartz, *Linear Operators; Part II: Spectral Theory*, Wiley, New York (1963).

[38]  M. S. P. Eastham, On the $L^2$ classification of fourth-order differential equations, *J. London Math. Soc. (2)* **3** (1971), 297–300.

[39]  M. S. P. Eastham, The limit–3 case of self–adjoint differential expressions of the fourth order with oscillating coefficients, *J. London Math. Soc. (2)* **8** (1974), 427–437.

[40]  M. S. P. Eastham, Self–adjoint differential equations with all solutions $L^2(0, \infty)$, *Differential Equations, Proceedings Uppsala 1977*, Almqvist & Wiksell, Stockholm, 52–61 (1977).

[41]  M. S. P. Eastham and C. G. M. Grudniewicz, Asymptotic theory and deficiency indices for fourth and higher order self-adjoint equations: a simplified approach, in: *Ordinary and Partial Differential Equations*, eds. W. N. Everitt and B. D. Sleeman, Lecture Notes in Math. Vol. **846**, Springer–Verlag, Berlin, 88–99 (1981).

[42]  M. S. P. Eastham and M. L. Thompson, On the limit–point, limit– circle classification of second ordinary differential equations, *Quart. J. Math. Oxford Ser. (2)* **24** (1973), 531–535.

[43]  J. Elias, On the solutions of $n$-th order differential equation in $L^2(0, \infty)$, in: *Qualitative Theory of Differential Equations Szeged (Hungary), 1979*, ed. M. Farkas, Colloquia Mathematica Societatis János Bolyai, Vol. **30**, North-Holland, Amsterdam, 181–191 (1981).

[44]  W. D. Evans and A. Zettl, Interval limit-point criteria for differential expressions and their powers, *J. London Math. Soc. (2)* **15** (1977), 119–133.

[45]  W. N. Everitt, On the limit–point classification of second order differential operators, *J. London Math. Soc.* **41** (1966), 531–534.

[46]  W. N. Everitt, On the limit–point classification of fourth–order differential equations, *J. London Math. Soc.* **44** (1969), 273–281.

[47]  W. N. Everitt, On the limit–circle classification of second–order differential expressions, *Quart. J. Math. Oxford Ser. (2)* **23** (1972), 193–196.

[48]  W. N. Everitt, On the strong limit–point condition of second–order differential expressions, *Proceedings of the International Conference on Differential Equations, Los Angeles*, **23**, Academic Press, New York, 287–307 (1974).

[49] W. N. Everitt, On the deficiency index problem for ordinary differential operators 1910–1976, *Differential Equations, Proceedings Uppsala 1977*, Almqvist & Wiksell, Stockholm, 62–81 (1977).

[50] W. N. Everitt and V. K. Kumar, On the Titchmarsh–Weyl theory of ordinary symmetric differential expressions I: the general theory, *New Archief voor Wiskunde (3)* **34** (1976), 1–48.

[51] W. N. Everitt and V. K. Kumar, On the Titchmarsh–Weyl theory of ordinary symmetric differential expressions II: the odd–order case, *New Archief voor Wiskunde (3)* **34** (1976), 109–145.

[52] M. V. Fedorjuk, Asymptotics of solutions of ordinary linear differential equations of $n$-th order, *Dokl. Akad. Nauk SSSR* **165** (1965), 777–779.

[53] M. V. Fedorjuk, Asymptotic method in the theory of one–dimensional singular differential operators, *Trudi Mosk. Mat. Obsch.* **15** (1966), 296–345. (English translation: *Trans. Moscow Math. Soc.* **15** (1966), 333–386.)

[54] M. V. Fedorjuk, *Asymptotic Analysis*, Springer-Verlag, New York, (1993).

[55] K. O. Friedrichs, Über die ausgezeichnete Randbedingung in der Spektraltheorie der halbbeschränkten gewöhnlichen Differentialoperatoren zweiter Ordnung, *Math. Ann.* **112** (1935), 1–23.

[56] V. Garbušin, Inequalities of the norms of a function and its derivatives in $L_p$ metric, *Mat. Zametki* **1** (1967), 291–298. (In Russian).

[57] I. M. Glazman, *Direct Methods of Qualitative Spectral Analysis of Singular Differential Operators*, Davey, Jerusalem, (1965).

[58] J. R. Graef, Limit circle type criteria for nonlinear differential equations, *Proc. Japan Acad.* **55** (1979), 49–52.

[59] J. R. Graef, Limit circle criteria and related properties for nonlinear equations, *J. Differential Equations* **35** (1980), 319–338.

[60] J. R. Graef, Limit circle type results for sublinear equations, *Pacific J. Math.* **104** (1983), 85–94.

[61] J. R. Graef, L. Hatvani, J. Karsai, and P. W. Spikes, Boundedness and asymptotic behavior of solutions of second order nonlinear differential equations, *Publ. Math. Debrecen* **36** (1989), 85–99.

[62] J. R. Graef and P. W. Spikes, Asymptotic behavior of solutions of a second order nonlinear differential equation, *J. Differential Equations* **17** (1975), 461–476.

[63] J. R. Graef and P. W. Spikes, Asymptotic properties of solutions of a second order nonlinear differential equation, *Publ. Math. Debrecen* **24** (1977), 39–51.

[64] J. R. Graef and P. W. Spikes, The limit point-limit circle problem for nonlinear equations, in: *Spectral Theory of Differential Operators*, North-Holland Mathematics Studies, Vol. **55**, North Holland, Amsterdam, 207–210 (1981).

[65] J. R. Graef and P. W. Spikes, On the nonlinear limit–point/limit–circle problem, *Nonlinear Anal.* **7** (1983), 851–871.

[66] J. R. Graef and P. W. Spikes, Some asymptotic properties of solutions of $(a(t)x')' - q(t)f(x) = r(t)$, in: *Differential Equations: Qualitative Theory (Szeged, 1984)*, Colloquia Mathematica Societatis János Bolyai, Vol. **47**, North-Holland, Amsterdam, 347–359 (1987).

[67] M. K. Grammatikopoulos and M. R. Kulenović, On the nonexistence of $L^2$–solutions of $n$-th order differential equations, *Proc. Edinburgh Math. Soc.* **24** (1981), 131–136.

[68] M. Greguš, *Third Order Linear Differential Equations*, D. Reidel Publ. Comp., Dordrecht, (1987).

[69] R. C. Grimmer and W. T. Patula, Nonoscillatory solutions of forced second-order linear equations, *J. Math. Anal. Appl.* **56** (1976), 452–459.

[70] T. G. Hallam, On the nonexistence of $L^p$ solutions of certain nonlinear differential equations, *Glasgow Math. J.* **8** (1967), 133–138.

[71] B. J. Harris, Limit–circle criteria for second-order differential expressions, *Quart. J. Math. Oxford (2)* **35** (1984), 415–427.

[72] P. Hartman, *Ordinary Differential Equations*, 2 Ed., Birkhäuser, Boston-Basel-Stuttgart, (1982).

[73] P. Hartman and A. Wintner, Criteria of non–degeneracy for the wave equation, *Amer. J. Math.* **70** (1948), 295–308.

[74] P. Hartman and A. Wintner, A criteria for the non–degeneracy of the wave equation, *Amer. J. Math.* **71** (1949), 206–213.

[75] D. B. Hinton, Limit point–limit circle criteria for $(py')' + qy = \lambda ky$, in: *Ordinary and Partial Differential Equations*, eds. B. D. Sleeman and I. M. Michael, Lecture Notes in Math. Vol. **415**, Springer-Verlag, New York, 173–183 (1974).

[76] R. M. Kauffman, On the limit-$n$ classification of ordinary differential operators with positive coefficients, *Proc. London Math. Soc.* **35** (1977), 496–526.

[77] R. M. Kauffman, T. T. Read, and A. Zettl, *The Deficiency Index Problem for Powers of Ordinary Differential Expressions*, Lecture Notes in Math. Vol. **621**, Springer-Verlag, New York (1977).

[78] I. T. Kiguradze, On the oscillation of solution of the equation $d^m u/dt^m + a(t)|u|^n \operatorname{sgn} u = 0$, *Mat. Sb.* **65** (1964), 172–187. (In Russian.)

[79] I. T. Kiguradze, *Some Singular Boundary-Value Problems for Ordinary Differential Equations*, University of Tbilisi (1975).

[80] I. T. Kiguradze and T. A. Chanturia, *Asymptotic Properties of Solutions of Nonautonomous Ordinary Differential Equations*, Kluwer, Dordrecht (1993).

[81] I. Knowles, The limit–point and limit–circle classification of the Sturm–Liouville operator $(py') + qy$, *Ph.D. thesis, Flinders University of South Australia*, (1972).

[82] I. Knowles, On a limit–circle criterion for second-order differential operators, *Quart. J. Math. Oxford Ser.* (2) **24** (1973), 451–455.

[83] I. Knowles, On second-order differential operators of limit circle type, in: *Ordinary and Partial Differential Equations*, eds. B. D. Sleeman and I. M. Michael, Lecture Notes in Math. Vol. **415**, Springer-Verlag, New York, 184–187 (1974).

[84] I. Knowles, Note on a limit–circle criterion, preprint.

[85] A. M. Krall, On the solutions of $(ry')' + qy = f$, *Monatsh. Math.* **80** (1975), 115–118.

[86] A. Kroopnick, $L^2$–solutions to $y'' + c(t)y' + a(t)b(y) = 0$, *Proc. Amer. Math. Soc.* **39** (1973), 217–218.

[87] A. J. Kroopnick, Note on bounded $L^p$–solutions of a generalized Liénard equation, *Pacific J. Math.* **94** (1981), 171–175.

[88] A. J. Kroopnick, Note on a bounded $L^p$–solution to $x'' - a(t)x^c = 0$, *J. Math. Anal. Appl.* **113** (1986), 451–453.

[89] M. K. Kwong, On the boundedness of solutions of second order differential equations in the limit circle case, *Proc. Amer. Math. Soc.* **52** (1975), 242–246.

[90] M. K. Kwong and A. Zettl, *Norm Inequalities for Derivatives and Differences*, Lecture Notes in Math. Vol. **1536**, Springer–Verlag, New York (1992).

[91] G. Ladas G., Connection between oscillation and spectrum for selfadjoint differential operators of order 2n, *Comm. Pure Appl. Math.* **XXII** (1969), 561–585.

[92] N. Levinson, Criteria for the limit–point cases for second order linear differential operators, *Časopis Pěst. Mat.* **74** (1949), 17–20.

[93] M. Marini and P. L. Zezza, On the asymptotic behavior of the solutions of a class of second-order linear differential equations, *J. Differential Equations* **28** (1978), 1–17.

[94] M. A. Naimark, *Linear Differential Operators, Part II*, George Harrap & CO., LTD, London, 1968.

[95] F. Neuman, On a problem of transformations between limit–circle and limit–point differential equations, *Proc. Roy. Soc. Edinburgh* **72A** (1973), 187–193.

[96] F. Neuman, *Global Properties of Linear Ordinary Differential Equations*, Academia, Praha (1991).

[97] R. B. Paris and A. D. Wood, On the $L^2$ nature of solutions of $n$-th order symmetric differential equations and McLeod's conjecture, *Proc. Roy. Soc. Edinburgh* **90A** (1981), 209–236.

[98] W. T. Patula and P. Waltman, Limit point classificaiton of second order linear differential equations, *J. London Math. Soc. (2)* **8** (1974), 209–216.

[99] W. T. Patula and J. S. W. Wong, An $L^p$-analogure of the Weyl alternative, *Math. Ann.* **197** (1972), 9–28.

[100] I. A. Pavljuk, Necessary and sufficient conditions for boundedness in the space $L^2(0, \infty)$ for solutions of a class of linear differential equations of second order, *Dopovidi Akad. Nauk. Ukrain. RSR* **1960** (1960), 156–158. (In Ukrainian).

[101] M. Ráb, Asymptotic formulas for the solution of linear differential equations of the second order, in: Differential Equations and Their Applications (Proc. Conf., Prague, 1962), Academia, Praha, 131–135 (1963).

[102] B. Schultze, On singular differential operators with positive coefficients, *Proc. Roy. Soc. Edinburgh* **120A** (1992), 361–365.

[103] D. B. Sears, On the solutions of a second order differential equation which are square integrable, *J. London Math. Soc.* **24** (1949), 207–215.

[104] P. W. Spikes, On the integrability of solutions of perturbed nonlinear differential equations, *Proc. Roy. Soc. Edinburgh Sect. A* **77** (1977), 309–318.

[105] P. W. Spikes, Criteria of limit circle type for nonlinear differential equations, *SIAM J. Math. Anal.* **10** (1979), 456–462.

[106] S. Staněk, Bounds for solutions of nonlinear differential equation of the third order, *Acta Univ. Palac. Olom. FRN Math.* XXVI **88** (1987), 47–55.

[107] L. Suyemoto and P. Waltman, Extension of a theorem of A. Winter, *Proc. Amer. Math. Soc.* **14** (1963), 970–971.

[108] M. Švec, On various properties of the solutions of third and fourth-order linear differential equations, in: Differential Equations and Their Applications (Proc. Conf., Prague, 1962), Academia, Praha, 187–198 (1963).

[109] E. C. Titchmarsh, *Eigenfunction Expansions Associated with Second-Order Differential Equations*, Part I, Oxford Univ. Press, Oxford (1962).

[110] E. C. Titchmarsh, On the uniqueness of the Green's function associated with a second-order differential equation, *Canad. J. Math.* **1** (1949), 191–198.

[111] P. W. Walker, Deficiency indices of fourth-order singular differential opeators, *J. Differential Equations* **9** (1971), 133–140.

[112] P. W. Walker, Asymptotics for a class of fourth order differential equations, *J. Differential Equations* **11** (1972), 321–334.

[113] J. Weidmann, *Spectral Theory of Ordinary Differential Operators*, Lecture Notes in Math. Vol. **1258**, Springer–Verlag, New York (1987).

[114] H. Weyl, Über gewöhnliche Differentialgleichungen mit Singularitäten und die zugehörige Entwicklung willkürlicher Funktionen, *Math. Ann.* **68** (1910), 220–269.

[115] A. Wintner, A criterion for the nonexistence of $(L^2)$-solutions of a nonoscillatory differential equation, *J. London Math. Soc.* **25** (1950), 347–351.

[116] J. S. W. Wong, Remark on a theorem of A. Wintner, *Enseignement Math. (2)* **13** (1967), 103–106.

[117] J. S. W. Wong, On a limit point criterion of Weyl, *J. London Math. Soc. (2)* **1** (1969), 35–36.

[118] J. S. W. Wong, Remarks on the limit–circle classification of second order differential operators, *Quart. J. Math. Oxford (2)* **24** (1973), 423–425.

[119] J. S. W. Wong, Square integrable solutions of $L^p$ peturbations of second order linear differential equations, in: *Ordinary and Partial Differential Equations*, eds. B. D. Sleeman and I. M. Michael, Lecture Notes in Math. Vol. **415**, Springer-Verlag, New York, 282–292 (1974).

[120] J. S. W. Wong, Square integrable solutions of perturbed linear differential equations, *Proc. Roy. Soc. Edinburgh Sect. A* **73** (1974–1975), 251–254.

[121] J. S. W. Wong and A. Zettl, On the limit point classification of second order differential equations, *Math. Z.* **132** (1973), 297–304.

# Author Index

# Subject Index